Cell Surface and Differentiation

Cell Surface and Differentiation

TAKASHI MURAMATSU
Professor,
Department of Biochemistry, Faculty of Medicine,
Kagoshima University, Japan

CHAPMAN AND HALL
LONDON • NEW YORK • TOKYO • MELBOURNE • MADRAS

UK	Chapman and Hall, 11 New Fetter Lane, London EC4P 4EE
USA	Chapman and Hall, 29 West 35th Street, New York NY10001
JAPAN	Chapman and Hall Japan, Thomson Publishing Japan, Hirakawacho Nemoto Building, 7F, 1-7-11 Hirakawa-cho, Chiyoda-ku, Tokyo 102
AUSTRALIA	Chapman and Hall Australia, Thomas Nelson Australia, 480 La Trobe Street, PO Box 4725, Melbourne 3000
INDIA	Chapman and Hall India, R. Sheshadri, 32 Second Main Road, CIT East, Madras 600 035

Typeset in Great Britain by EJS Chemical Composition,
Midsomer Norton, Bath, Avon
Printed in Great Britain by St Edmundsbury Press Ltd,
Bury St Edmunds, Suffolk

ISBN 0 412 30850 9

British Library Cataloguing in Publication Data

Muramatsu, Takashi
Cell surface and differentiation.
1. Mammals. cells. Membranes
I. Title
599.0875

ISBN 0 412 30850 9

Library of Congress Cataloging-in-Publication Data available

Contents

Preface

Cell surface membranes contain diverse kinds of molecules, notably proteins. They play critical roles in the regulation of cellular activities such as growth and differentiation. Knowledge of the structure and function of cell surface molecules has increased enormously in recent years, partly because of recombinant DNA techniques.

This book describes the current state of a fast growing research area, namely the molecular biology of the cell surface with respect to cell differentiation. The role of cell surface molecules in differentiation is the central subject of this book, but I have also dealt with cell surface markers, which are useful in monitoring differentiation, and cell adhesion molecules, which influence differentiation. Since the book is multidisciplinary in nature and the readers are expected to include students and graduate students, I have included two introductory chapters in order to familiarize readers with current developmental biology and molecular studies of the cell surface. Although the scope covered by this book is large, I tried to keep it concise, so that an interested reader can read through it in a few days.

I am grateful to Dr Paul H. Atkinson, who offered helpful criticisms and suggestions regarding the content, to Dr Susan Hemmings of Chapman and Hall for hearty co-operation and to Miss Kumiko Sato for all the secretarial work needed to prepare the manuscript.

1 Biological systems

A fertilized egg gives rise to many different cell types, such as nerve cells, myoblasts, cartilage cells and endothelial cells. The whole or a part of the process where a cell yields two or more cell types is called differentiation [1]. The elucidation of the mechanism of differentiation is one of the most fascinating studies in biological sciences. Differentiation is most often observed during embryogenesis; but even in adulthood certain cells, such as blood cells and the epithelial cells of the small intestine, differentiate continuously from stem cells. Important concepts in classical embryology were formulated from studies using the sea urchin, newt and chicken. The processes of development in these animals are relatively easy to observe. Furthermore, the large size of the newt and chick embryos have allowed experimental manipulation of the developmental processes by surgical procedure. However, the current preferred systems study the molecular mechanism of embryogenesis, and are usually the fruit fly, nematode and mouse, although the use of amphibians is being revived in exciting experiments in frog embryos. Undoubtedly, organisms in which a genetic approach is possible are much preferred. This introductory chapter deals with the biological systems used frequently in developmental biology and some of the concepts of the discipline.

1.1 NEMATODES AND CELL LINEAGE

The formation of a differentiated cell from a fertilized egg consists of many steps. Several intermediate cells arise before the final appearance of a differentiated cell, and the pathway of differentiation is called cell lineage. A description of cell lineage is essential for a precise understanding of differentiation. In each step of cell lineage, namely from one cell to the next cell, the following questions must be answered:

1. Is the differentiation pathway one-way, bi-directional or multi-directional?
2. Does the step proceed autonomously or is it influenced by an external signal?
3. If influenced, what is the external signal?

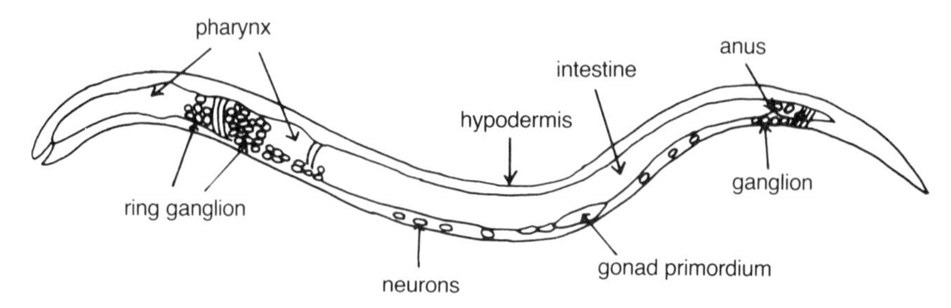

Figure 1.1 A larva of *Caenorhabditis elegans* (based on [2]).

The cell lineage occurring to form a complete organism has been determined only in one organism, the nematode *Caenorhabditis elegans* (Figure 1.1) [2–4]. This worm develops quickly – the 560-cell embryo hatches 14 hours after fertilization. About a day after hatching, it becomes a mature worm which is 1 mm in length and has 959 somatic cells. The cell lineage was determined by observing the fate of individual cells by a microscope, and an outline is described in Figure 1.2. After the first division, a fertilized egg produces two asymmetrical cells designated AB and P_1. The majority of the progeny of AB go on to form dermis and nerve cells. From P_1, EMS and P_2 cells are formed, and EMS produces MS and E. The progeny of E are destined to be intestinal cells, while MS produces many cell types as shown in Figure 1.3. P_2 forms P_3 and C; P_3 produces P_4 and D.

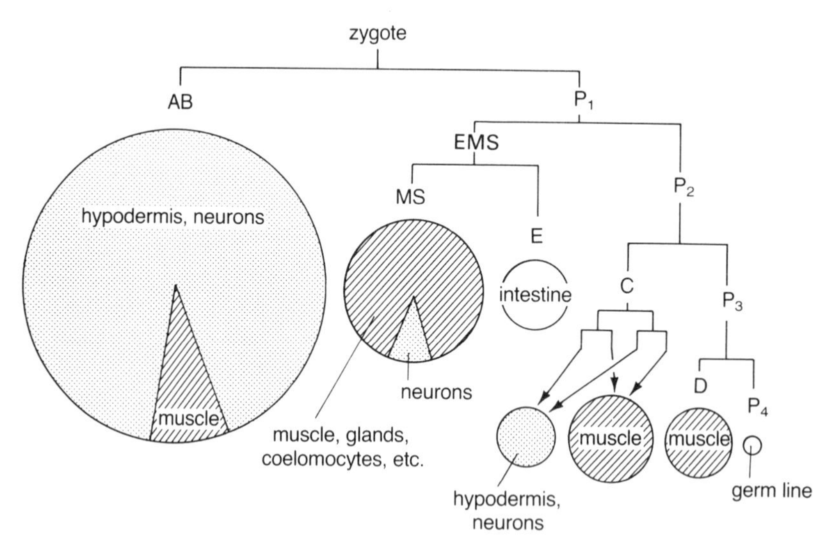

Figure 1.2 Summary of cell-types derived from blastomeres produced by early cleavage. Areas of circles and sectors are proportional to number of cells (cited from [2]). © Academic Press.

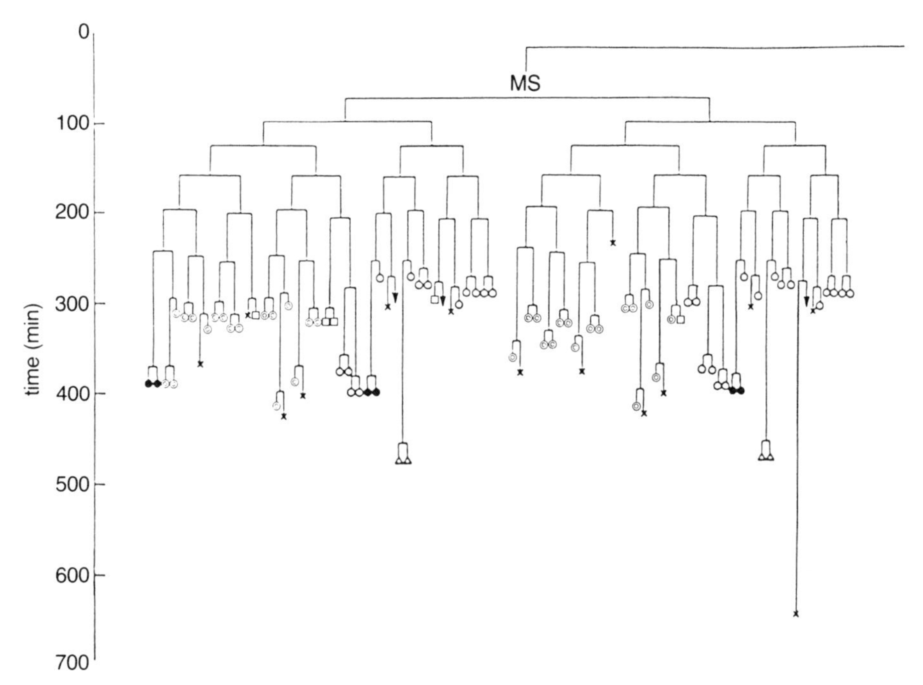

Figure 1.3 A part of cell lineage of *C. elegans* (progeny of MS cell). ● neuron; ◎ pharynx cells; ○ muscle cells; △coelomocytes; □ other cells, unidentified cells; × death; ➤ cells that divide in larval development (cited from [2]).

P_4 forms only germ cells, while D is devoted to form only muscle. It is notable that the cell lineage which forms even a tiny worm is very complex.

When a cell in the intermediate stage of differentiation was selectively killed by a lazor beam, the developmental fate of other cells was largely unaffected. Thus, the cell lineage of *C. elegans* was considered to be 'rigid'. Fertilized eggs are often classified into mosaic and regulative types [1]. In the mosaic type the fate of blastomeres created by initial cell divisions is largely predetermined, while in the regulative type the fates of the initial blastomeres are not determined. In view of the rigid lineage, the egg of *C. elegans* is a typical example of a mosaic type. The observation that cell fate is predetermined for each blastomere of a mosaic egg implies that a certain cytoplasmic determinant of cell differentiation is present in the egg cytoplasm. The unequal distribution of the determinant which results from cell division plays an important role in the embryogenesis of mosaic eggs.

However, even in *C. elegans*, the differentiation of certain cells does not proceed independently of interactions with other cells. A typical example is muscle formation [5]. In the normal embryo, the progenies of both the AB blastomere and the P_1 blastomere produce muscle. When the P_1 blastomere is isolated and is allowed to develop in the absence of the AB blastomere, the progeny cells are able to produce muscle. However, progenies of isolated AB blastomere cannot produce muscle. Further analysis has revealed that interaction with the progeny of the EMS cell is required for the progeny of the AB blastomere to produce muscle. Furthermore, when the position of ABa and ABp, which are daughter cells of AB, are exchanged, the progenies of ABa develop as if they are the progenies of ABp cell and vice versa [5].

1.2 MOUSE EMBRYOGENESIS AND ITS MANIPULATION

The mouse is most often chosen in the study of the mechanism of mammalian embryogenesis because the duration of its embryogenesis is short, namely 19–20 days, and the small animal is easy to handle. In addition, there are many useful inbred strains available, especially some mutant strains which are abnormal in certain developmental processes.

The earlier stages of mouse embryogenesis are summarized in Figure 1.4. The fertilized egg divides about 18 hours after fertilization, and subsequent

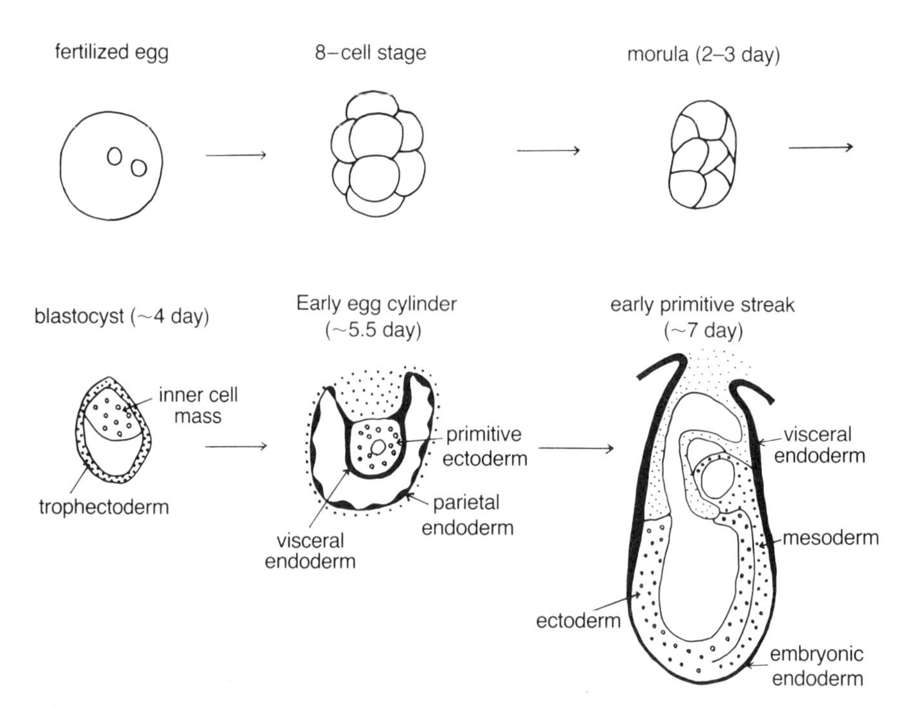

Figure 1.4 Early embryogenesis of the mouse (partly based on [6–8]).

divisions take place every 12 hours during the preimplantation stage. Until the 8-cell stage, each blastomere in an embryo looks identical. At the late 8-cell stage, cell adhesion between blastomeres increases, and the embryo becomes more compact; this phenomenon is called compaction. The embryo at the 16-cell stage is called a morula. At the 32-cell stage two cell groups – externally located trophectoderm and internally located inner cell mass – can be distinguished clearly. The embryo develops a blastocoel and is called a blastocyst. Trophectoderm cells constitute placenta and parietal yolk sac, and the fetus develops from the inner cell mass. At the late blastocyst stage, the embryo implants into the uterine wall.

During implantation, a layer of primitive endoderm cells develops on the surface of the inner cell mass, and the inner cells become primitive ectoderm cells. Primitive endoderm cells differentiate into two types of extraembryonic endoderm cells, namely visceral endoderm cells, which cover the embryo proper, and parietal endoderm cells, which underlay trophectoderm and secrete components of a kind of basement membrane called Reichert membrane. Around day 7 and 8 of embryogenesis, ectoderm cells, mesoderm cells and endoderm cells differentiate from primitive ectoderm cells. On Day 9, the embryo already shows the characteristic features of a vertebrate embryo. Cell layers then interact with each other and form organs.

During embryogenesis many cells migrate for long distances. For example, precursor cells to blood cells migrate from the yolk sac to the thymus, liver and bone marrow; and primordial germ cells, which are destined to be germ cells, migrate from the base of the allantois to the hind gut, and then to the gonads. A detailed description of mouse embryogenesis is found in references [6], [7] and [8].

Mouse embryogenesis is a typical example of the regulative type: When a blastomere at the 2-cell stage or 4-cell stage was destroyed by a fine needle, other blastomeres compensated for the destruction, and normal embryos developed from the treated ones [9]. A single 8-cell blastomere was also shown to be developmentally totipotent, namely it constituted both embryonic and extraembryonic tissues. Positional information appears to determine the fate of a blastomere at this stage and one of the two differentiation pathways (trophectoderm cells and cells of the inner cell mass) is selected. A blastomere located externally at the 16-cell stage differentiates to trophectoderm cells, while a blastomere located internally differentiates to inner cell mass cells. Indeed, two different cell populations are found at the 16-cell stage; externally located large cells and internally located small cells. The origin of the two distinct cell populations can be traced back to the late 8-cell stage [10]. At that time, asymmetry can be detected on the plasma membrane. In the membrane facing the external surface, microvilli develop and several cell-surface proteins are concentrated on this side. This phenomenon is called polarization. When cell division occurs parallel to the polarized area, two different cells, namely a polarized large cell and an apolar small cell are generated (Figure 1.5). These observations might indicate that a polarized distribution of cell surface components is associated with the determination of cell fate. However, aggregates composed of a pure population of the large cells as well as those composed of a pure population of small cells can yield normal embryos [11]. Cells of the 16-cell stage might have selected a developmental programme (to yield either trophectoderm cells or inner cell mass cells) but the decision can be altered if there is a drastic change in the environment.

Positional information also appears to be of critical importance in the differentiation of inner cell mass cells into primitive ectoderm cells and primitive endoderm cells. Cells exposed at the surface of the inner cell mass are likely to be triggered to differentiate into primitive endoderm cells.

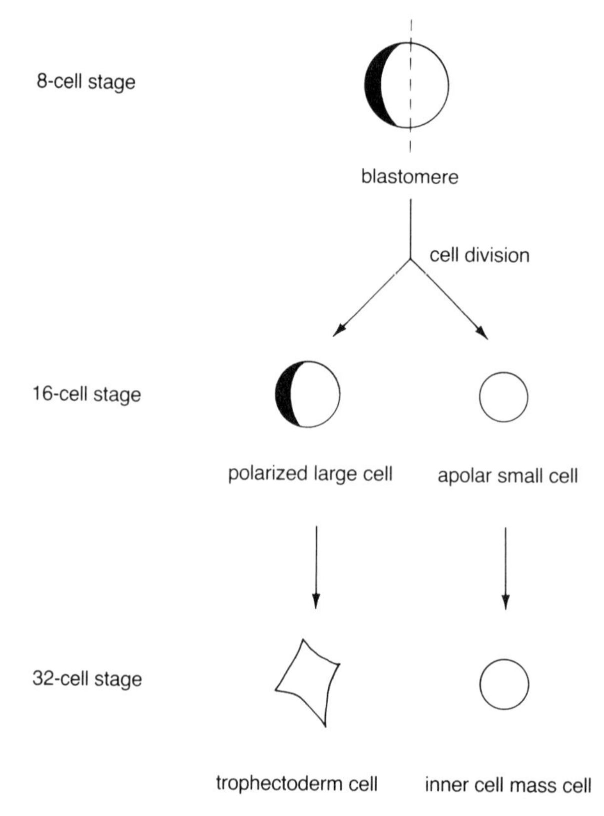

Figure 1.5 Polarization of cell surface molecules and differentiation of trophectoderm cells and cells of inner cell mass.

The flexibility in mouse embryogenesis is also illustrated by the formation of chimeric mice. When two morulae are aggregated, embryos with twice the number of cells adjust themselves and yield mice of normal size and characteristics [12]. The chimeric nature of the progeny can be demonstrated clearly by the mosaicism of the coat colour observed when embryos are used from mice which differ genetically in their coat colour.

The mouse is frequently used in experiments to introduce and express foreign genes in a multicellular organism. For that purpose, exogenous DNA is microinjected into the male pronucleus of the fertilized egg [13]. In the fertilized egg, there are two pronuclei, one derived from the sperm (male pronucleus) and the other from the egg (female pronucleus). Genes are introduced into the male pronucleus, which is bigger than the female one and is more suitable for injection experiments. The fertilized egg with injected DNA is either directly grafted into the oviduct of a foster mouse or cultured *in vitro* to the blastocyst stage and grafted to the uterus of the foster mother. Mice which are subsequently born with integrated foreign DNA are called transgenic mice. Thus, introduction of a growth hormone gene with a strong promoter resulted in the production of a super mouse which had around 1.5 times the body weight of ordinary mice [14]. This result demonstrated the effectiveness of transgenic techniques.

Microinjection is not the only way to generate transgenic mice. Even, the usage of sperm as a vector of foreign DNA was once proposed [15], but the result has not been confirmed so far.

The ability to produce transgenic mice is expected to be helpful in elucidating the role of various molecules during differentiation and development. The most successful approach so far is to introduce a gene with a strengthened promoter, so that the gene becomes expressed at an inappropriate time and place. Abnormalities observed in the resulting transgenic mice give information on the function of the gene product [16]. Inhibition of gene function by the introduction of an artifical gene that produces anti-sense RNA of the target gene is also a promising approach. Indeed, a transgenic mouse with the abnormal nervous behaviour characteristic of shiverer disease was produced by introducing an anti-sense gene of myelin basic protein [17]. Homologous recombination to delete a target gene may also be a technique with general applications, although the technique is still under development [18]. Homologous recombination is performed in embryonic stem cells (ES cells) (section 1.5) to delete a gene in a chromosome, and chimeric mice are produced by injecting the ES cell into blastocysts. By mating the resulting mice, a mouse lacking the target gene both in the maternally derived chromosome and in the paternally derived chromosome is expected to arise.

Nuclear transplantation experiments can also be performed in mouse embryos. Male pronuclei and female pronuclei are removed from fertilized eggs with the aid of micropipets, and foreign nuclei are introduced by a fusion technique [19]. In such a way, a nucleus from a 16-cell embryo was found to direct the production of a mouse in the cytoplasm of a fertilized egg. However, it became difficult to form a mouse from nuclei from embryos of more advanced stages. It is known that a feeding tadpole can be formed when the nucleus of an unfertilized frog egg is replaced by a nucleus from adult frog cells such as keratynocytes. Thus, the developmental potential of nuclei appears to be profoundly different between the mouse and the frog. Even more surprising is the fact that in an egg whose pronuclei are reconstituted by transplantation, both the male pronucleus and female pronucleus are required to form a mouse – two male pronuclei or two female pronuclei are ineffective [20]. Therefore, it has been concluded that DNA (or chromosomes) in the sperm and the egg are modified in different manners and both types of DNA (or chromosomes) are needed to accomplish embryogenesis.

1.3 EARLY EMBRYOGENESIS OF AMPHIBIANS

Amphibian embryos have been the subject of experimental embryology for a long time, and are still helpful in studying the mechanism of the early stages of cell differentiation. When the early embryo of a newt was lassoed by a hair in the plane where the first cleavage furrow was expected, the two separated portions developed normally and formed twin newts [1]. This result thus clearly indicated the flexibility of the newt embryo. However, when the embryo was divided perpendiculary to the plane of the first cleavage, one half of the embryo formed the intact newt, while the other half yielded just a cell aggregate. Further studies by Hans Spemann have led to a conclusion that embryos with a middle portion called a grey crescent, which is seen in the region opposite the sperm-entry site, can develop into normal newts [1]. The former way of embryo cleavage usually resulted in a partition of the grey crescent into two portions. Therefore, an unequal distribution of a cytoplasmic component in the grey crescent region plays a decisive role in embryogenesis. Furthermore, Spemann and Mangold have suggested an inductive role for the dorsal lip, which is derived from the grey crescent. After gastrulation, the dorsal lip becomes part of the mesoderm, which is known as the 'organizer'. When the ectoderm was underlayed with the organizer, it developed into the neural tube. Otherwise, the ectoderm developed into the epidermis. Thus, the organizer apparently determines the fate of ectoderm cells. Various attempts have been made to isolate the chemical entity of the organizer but these have been unsuccessful. Subsequently, apparently non-specific stimulation, such as an alteration of the pH or ionic strength was found to induce neural differentiation in the ectoderm. Whether the signal delivered by the organizer is also non-specific, or it is specific and the non-specific stimulation just mimics or by-passes the action of the organizer is not known.

Recent studies have been conducted using the frog *Xenopus laevis*, which is easier to handle and has been yielding important results [21, 22]. When a *Xenopus* embryo was cut horizontally, and the separated pieces were cultured independently, the animal cap developed into the ectoderm and the vegetal portion into the endoderm, while no mesoderm was formed. When the two pieces were placed in contact, the mesoderm differentiated in the lower part of the animal cap. Thus, the vegetal portion has the capacity to induce mesoderm in ectoderm. Further studies recently carried out revealed the importance of growth factors in the induction process (Chapter 4).

As mentioned above, amphibian embryogenesis is a suitable system to show how cytoplasmic information controls embryogenesis. The asymmetry of the animal pole and the vegetal pole is sufficient to yield the ectoderm and the endoderm, and the interaction of the two embryonic portions yields the mesoderm. The dorsal–ventral axis of the embryo is determined by the point of sperm entry.

1.4 DEVELOPMENTAL GENETICS OF *DROSOPHILA*

The fruit fly *Drosophila melanogaster* has been used frequently in the molecular analysis of development [4, 23]. The adult fly is formed from a fertilized egg in only 9 days. In the giant polytene chromosomes found in the salivary glands of the larvae, about one thousand DNA strands and proteins assemble side by side and form characteristic and reproducible banding patterns. There are about 5 000 bands, and the average of each visible band is about 25 kb, and the size corresponds roughly to a gene. When a gene is mutated, especially deleted, one of the bands is usually altered. With this information, together with the result of crossing over experiments, the chromosomal position of the mutated gene can be determined accurately. Many mutants with certain abnormalities in developmental programs have been isolated using this organism. The advancement of recombinant DNA technology has made it possible to clone the gene defining the mutated phenotype by microdissecting the polytene chromosomes or by chromosome walking from the nearest known DNA segments. Chromosome walking is a procedure used to clone a gene, using a DNA probe whose sequence is located near to the desired gene. The identity of the cloned gene can be confirmed by rescuing the mutant phenotype through P-element-mediated transfer of the wild-type gene. Using these approaches, a number of genes specifying developmental steps have been cloned and sequenced. The structure of gene products predicted by the sequence data is providing new and deep insights into the mechanism of differentiation and development.

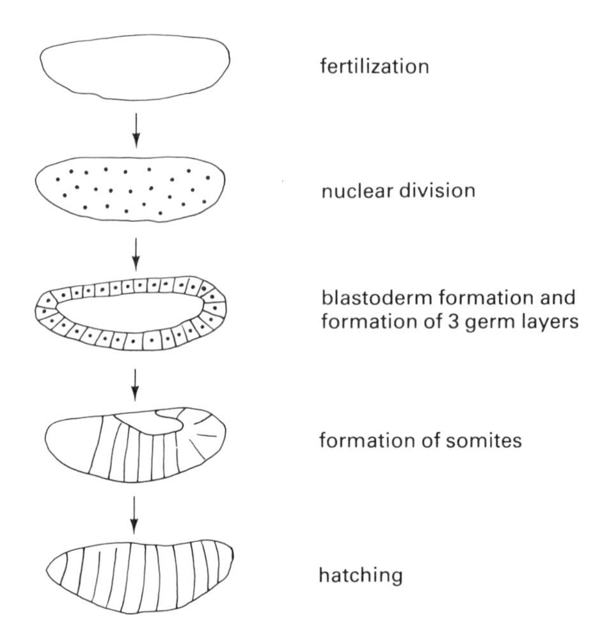

Figure 1.6 An outline of the embryogenesis of *D. melanogaster*.

The embryogenesis of *D. melanogaster* is illustrated in Figure 1.6. After fertilization, nuclear division occurs rapidly. However cell membranes are not formed at the initial stage of embryogenesis. About 1½ hours after the egg is laid, when the nuclei have divided 9 times, nuclei migrate to the surface of the egg. Then the nuclei divide another 4 times, and cell membranes are formed suddenly, resulting in the formation of a cellular blastoderm. At this stage, two embryonic axes are established; they are the anterior–posterior axis and the dorsal–ventral axis. Substances in the egg cytoplasm play important roles in establishing the embryonic axes: some of the maternal factors are asymmetrically distributed in the embryo and this asymmetry contributes to the establishment of the embryonic axes (Chapter 7). Shortly after the emergence of cellular blastoderm three germ layers are formed, and gastrulation follows. The embryo is then divided into segments, and the embryo then hatches – a day after fertilization.

1.5 DIFFERENTIATION OF TERATOCARCINOMA STEM CELLS

Cells which can differentiate *in vitro* are invaluable in studying the molecular mechanism of cell differentiation. Teratocarcinoma stem cells and intermediate cells of blood cell differentiation are important for this purpose. A teratocarcinoma is a tumour composed of cells of all three germ layers and of malignant stem cells [24]. The stem cells, called embryonal carcinoma cells (EC cells), resemble multipotential cells of early embryos, and have been used as a model for these cells. Indeed, upon injection into the inner cell mass of blastocysts, some EC cells differentiate normally and participate in the formation of tissues in normal mice. When EC cells in a tumour disappear as a result of differentiation, the tumour becomes benign. Benign tumours and teratocarcinomas are collectively called teratomas.

Teratomas arise spontaneously in the gonads of some inbred mice. They are believed to be derived from germ cells (or their precursors) which for some reason initiate monogametic embryogenesis: the multipotential cells of the abberantly formed embryos appear to become EC cells. Actually, early mouse embryos grafted into extra-uterine sites frequently develop teratomas [25]. Many cell lines of EC cells have been established from teratocarcinomas (Figure 1.7). Furthermore, cell lines can be established directly from mutlipotential cells of early mouse embryos by culturing blastocysts *in vitro*. The cells thus established are called embryonic stem cells (ES cells), and have properties similar to EC cells. However, the capacity for differentiation of ES cells is superior to that of EC cells, notably with respect to the formation of chimeric mice upon blastocyst injection [26].

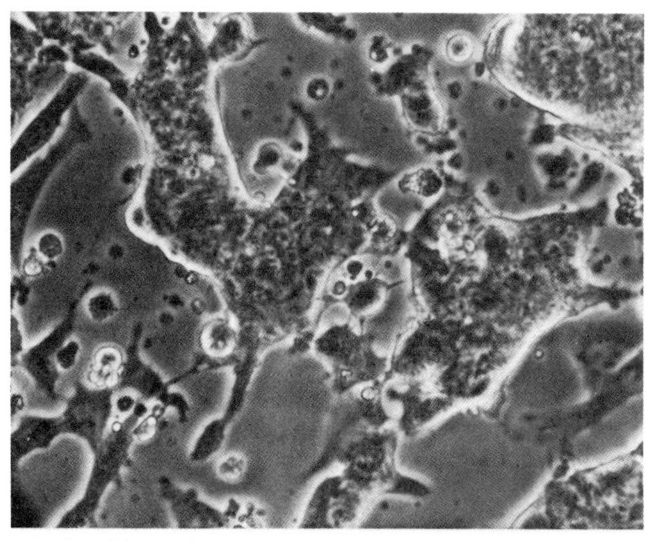

Figure 1.7 A clonal line of EC cells (HM-1 cells). Many cells adhere and form aggregates.

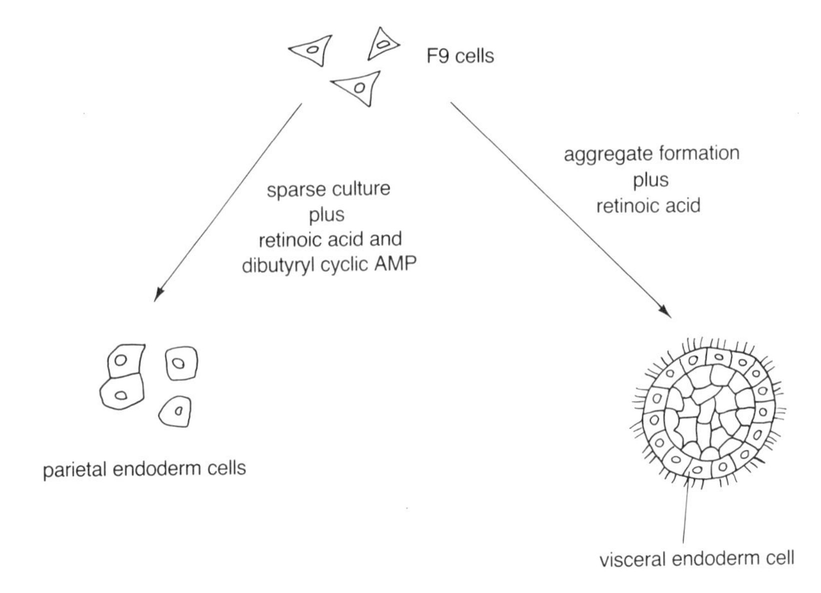

Figure 1.8 Bidirectional differentiation of F9 EC cells.

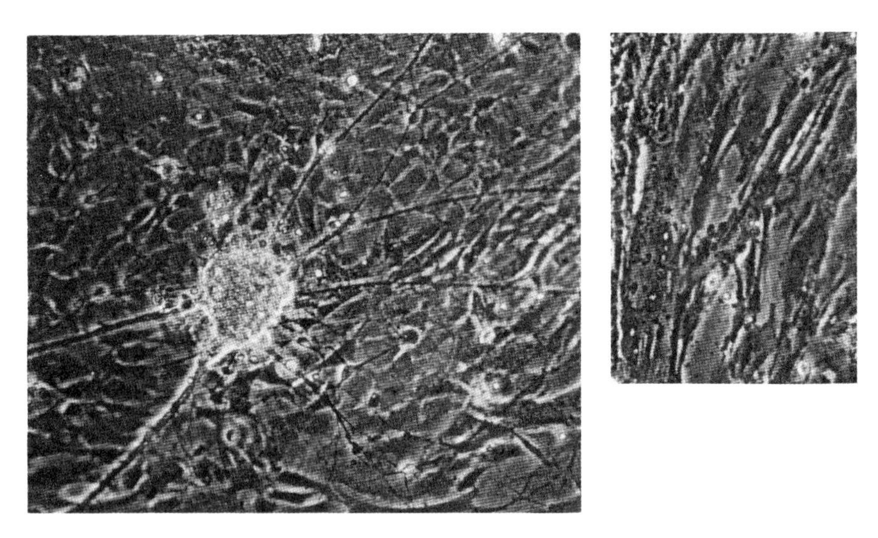

Figure 1.9 Cells differentiated from HM-1 cells. Left: a portion rich in neurons; right: myoblasts (reprinted from [31]). © Academic Press.

Table 1.1 Manipulation of differentiation of HM-1 EC cells

Conditions	Aggregate formation	Concentration of retinoic acid	Duration of retinoic acid treatment (days)	β-Mercapto-ethanol	Cells differentiated
1	+	10^{-6} M	2	+	Myoblasts Nerve cells Extraembryonic endoderm cells
2	+	10^{-7} M	2	+	Myoblasts Extraembryonic endoderm cells
3	+	10^{-6} M	2	–	Myoblasts Extraembryonic endoderm cells
4	+	10^{-6} M	4 or more	+	Myoblasts Nerve cells
5	–	10^{-6} M	14	+	Myoblasts

Under conditions 1–4, differentiation is almost complete in 8 days.
Data taken from [31].

EC cells and ES cells can be induced to differentiate *in vitro*. Aggregate formation is sufficient for the induction of differentiation in ES cells and some EC cells. As an example, aggregated PCC3 EC cells yield many cell types such as myoblasts, nerve cells, glia cells, chondrocytes, melanocytes, adipocytes, parietal endoderm cells and visceral endoderm cells [27].

Retinoic acid is a potent inducer of differentiation in EC cells [28]. When aggregates of F9 EC cells are treated with retinoic acid, the external cells differentiate into visceral endoderm cells [29]. When F9 cells are sparsely grown in tissue culture dishes and treated with retinoic acid and dibutyryl cyclic AMP, the cells differentiate into parietal endoderm cells (Figure 1.8).

Although F9 cells differentiate only to extraembryonic endoderm cells upon retinoic acid treatment, some EC cells such as P19 and HM-1, under

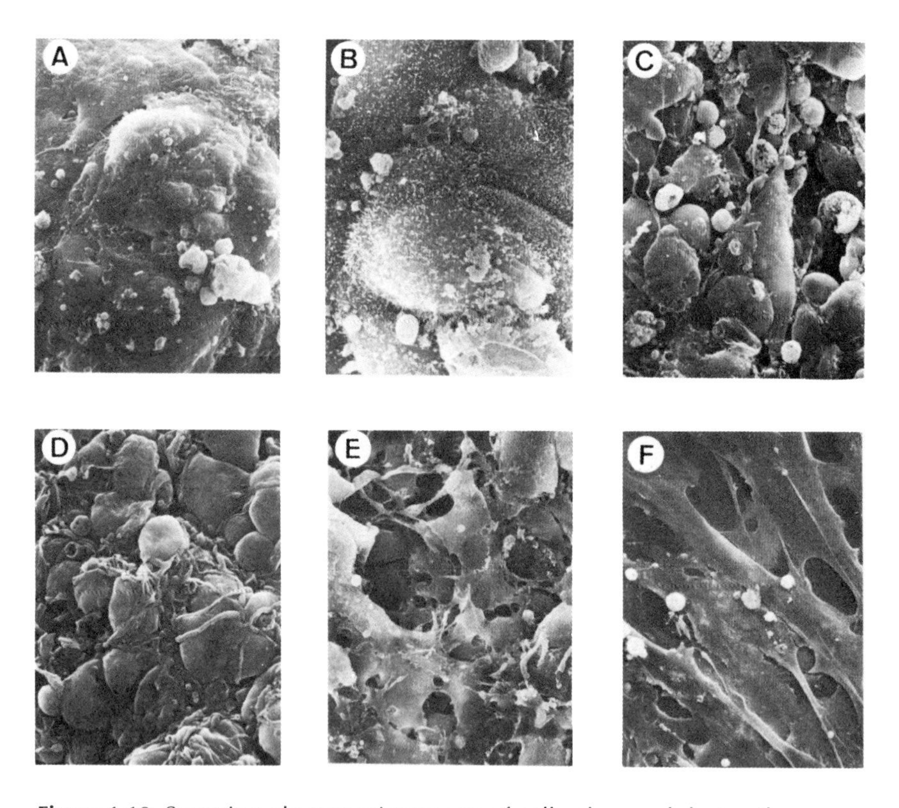

Figure 1.10 Scanning electronmicroscopy of cells observed during the course of retinoic acid-induced differentiation of HM-1 EC cells. A: cells with the morphology of undifferentiated EC cells; B: primitive endoderm cells; C, D: presumptive precursors to nerve cells; E, F: presumptive precursors to myoblasts (reprinted from [31]). © Academic Press.

corresponding treatment, differentiate to cells of the three germ layers (Figure 1.9). Furthermore, the concentration of retinoic acid has been found to alter the direction of differentiation of P19 cells. In 10^{-8} M retinoic acid, extraembryonic endoderm cells and cardiac myoblasts differentiate well, while a higher concentration of retinoic acid is required for nerve differentiation [30]. On the other hand, prolonged treatment with 10^{-6} M retinoic acid inhibits differentiation of extraembryonic endoderm cells from HM-1 EC cells [31]. The procedures used to manipulate the differentiation of HM-1 cells are summarized in Table. 1.1.

Such EC cell differentiation systems are suitable for analysing the mechanisms controlling the direction of differentiation and for elucidating the properties of cells at the intermediate stages of differentiation. The morphological features of intermediate cells observed during the differentiation of HM-1 cells are shown in Figure 1.10.

1.6 BLOOD CELL DIFFERENTIATION

Blood contains several kinds of cells, namely erythrocytes, granulocytes, monocytes and lymphocytes. All blood cells are believed to be derived from a common stem cell which is present in the bone marrow (Figure 1.11).

The following experiment revealed the presence of a stem cell involved in blood cell differentiation: When mice were irradiated by γ-rays, cells with high proliferating activity were killed. The treated mice could not form blood cells and eventually died. If bone marrow cells from mice of the same inbred strain were grafted into the irradiated mice, the grafted bone marrow cells grew in the spleen of the recipient mice, differentiated into blood cells and rescued the recipient. Grey nodules of about 1 mm diameter were detected in the spleen of the recipient mice. The nodule was composed of a

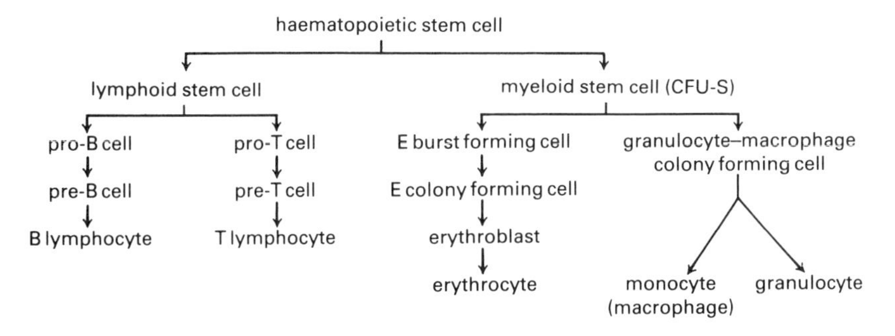

Figure 1.11 Cell lineage in the differentiation of blood cells.

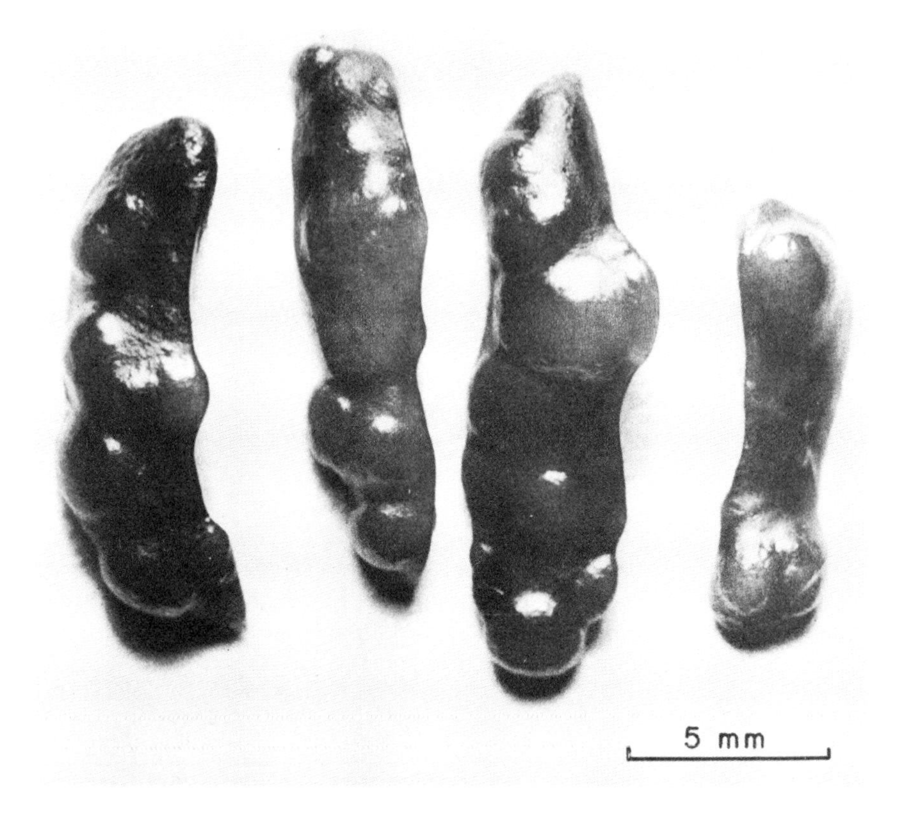

Figure 1.12 Spleen colony formed by myeloid stem cells (CFU-S) (reprinted from [32]). © Academic Press.

selection of hematopoietic cells including erythroblasts, and was considered to be derived from a single cell – now called the myeloid stem cell [32] (Figure 1.12).

Some stages of blood cell differentiation can be performed *in vitro*. Colonies composed of erythroblasts were detected when liver cells of mouse embryos were dissociated by trypsin digestion and cultured in a plasma clot with erythropoietin [33] (Figure 1.13). These colonies are thought to be derived from an intermediate cell named E rosette forming cell. When bone marrow cells were similarly cultured in higher concentrations of erythropoietin, colonies of E-rosette forming cells appeared and disappeared in 7 days [34]. Colonies composed of much larger numbers of cells were then detected. The colonies are thought to be formed by an

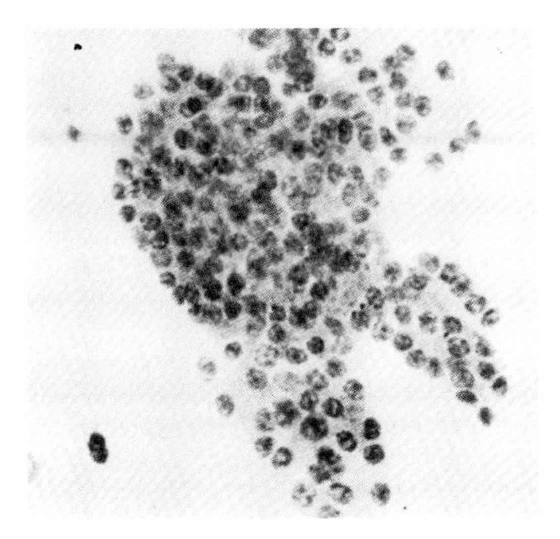

Figure 1.13 Colony formed *in vitro* by E-colony forming cells (reprinted from [33]).

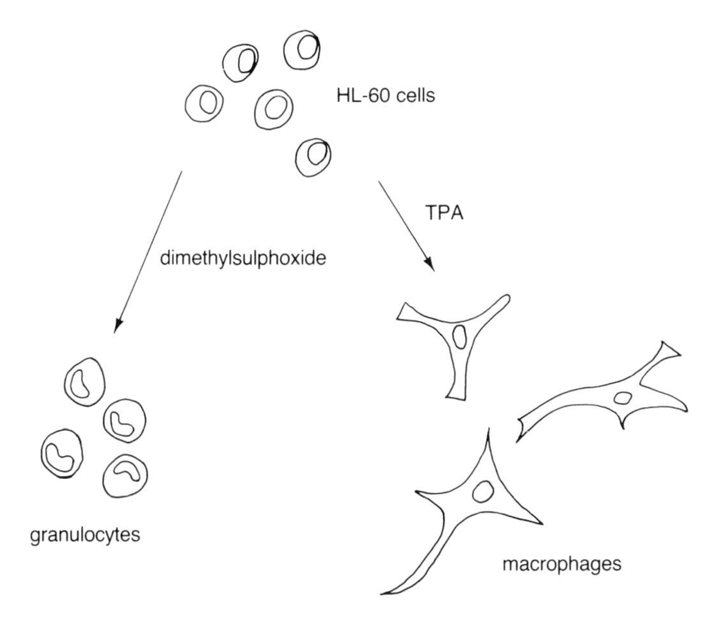

Figure 1.14 Bidirectional differentiation of HL-60 leukemia cells.

ancestor cell of E-rosette forming cells, namely an E burst forming cell (Figure 1.11).

Certain leukemia cells are regarded as cells arrested in blood cell differentiation at an intermediate stage. Using such leukemia cells, a step of blood cell differentiation can be analysed starting from a homogeneous population of intermediate cells. Friend leukemia cells and HL-60 leukemia cells are typical examples.

Friend leukemia cells were isolated from mice with Friend leukemia. After treatment with dimethylsulphoxide, the cells were able to differentiate along the erythroblast pathways to cells with hemoglobin synthesizing activity [35].

HL-60 cells were isolated from a patient with promyelocytic leukemia. When the cells were treated with dimethylsulphoxide, they differentiated to granulocytes [36]. On the other hand, the cells treated with a phorbol ester (TPA) differentiated to macrophages (Figure 1.14) [37]. Thus, these cells are also very useful for analysing the selection mechanism involved in controlling the direction of differentiation.

1.7 REGULATION OF DIFFERENTIATION DURING EMBRYOGENESIS

Although the mode of embryogenesis in the organisms mentioned above (the mouse, nematode, newt, frog and fruit fly) are quite different, the following two conclusions can be applied to all these and to many other organisms:

1. Differentiation consists of both autonomous cell steps and steps affected by information external to the cell. Even in mosaic eggs such as *C. elegans*, certain steps are influenced by external information.
2. The initial factor responsible for cell differentiation is believed to be the asymmetrical distribution of certain cellular components [38].

In *Drosophila*, *C. elegans* and amphibians, maternal substances in the egg cytoplasm are the key materials. In mouse embryos, positional information appears to cause the asymmetrical distribution. A critical question concerns the nature of asymmetrically distributed substances. In the cases of *Drosophila* and *Xenopus*, some of the maternal substances distributed asymmetrically are messenger RNAs which encode growth factors and DNA binding proteins (Chapters 4 and 7).

Although the initial signal triggering differentiation appears to be the unequal distribution of cellular material, transcriptional control is the key to the mechanism of differentiation. This is because genetic information stored in DNA is largely unchanged before and after differentiation. Earlier studies have shown that the content of DNA among cells of different phenotypes is invariable with a given organism. Using cloned genes as probes, more definitive evidence can be obtained. For example, the structure of silk fibroin gene is identical in the silk worm synthesizing fibroin to the gene in the one unable to synthesize it [39]. Nuclear transplantation experiments mentioned before have demonstrated that a differentiated frog cell contains all the information necessary to develop from an unfertilized egg into a feeding tadpole [38].

The essential point of transcriptional control is regulation by DNA binding proteins. In this respect, the control mechanism in eukaryotic cells appears not to be very different from the mechanism of transcriptional control found in prokaryotic cells. A typical example of the latter case is the control of β-galactosidase synthesis in *E. coli*, where repressor – a kind of DNA binding protein – negatively regulates the gene expression [40]. In eukaryotic cells, steroid hormone receptors and many other DNA binding proteins have been identified recently and some of them have been verified as participating in transcriptional control.

In addition to control by DNA binding proteins, changes in the chemical structure of a gene is also involved in the regulation of gene expression in eukaryotic cells. A rearrangement of immunoglobulin genes during the differentiation of lymphocytes is well established, and the rearrangement is required for gene expression [41]. Ribosomal RNA gene is amplified during the formation of oocytes in order to fulfil the requirement of a large number of ribosomes [42]. Furthermore, methylation of DNA is believed to be involved in control of gene expression [43]. Since the differentiated phenotypes of an eukaryotic cell are stable, it is possible that certain chemical modification of DNA is involved in maintenance of the stability.

REFERENCES

1. Gilbert, S.F. (1988) *Developmental Biology*, Sinauer Associates, Sunderland, Mass
2. Sulston, J.E., Schierenberg, E., White, J.G. and Thomson, J.N. (1983) *Dev. Biol.*, **100**, 64–119
3. Kenyon, C. (1988) Science, **240**, 1448–52
4. Alberts, B., Bray, D., Lewis, J., Raff, M., Roberts, K. and Watson, J.D. (1989) *Molecular Biology of the Cell*, Garland Publishing, New York
5. Priess, J.R. and Thompson, J.N. (1987) *Cell*, **48**, 241–50
6. Hogan, B. (1986) *Manipulating the mouse embryo: A laboratory manual*, Cold Spring Harbor Laboratory, New York
7. Theiler, K. (1972) *The House Mouse*, Springer-Verlag, Berlin
8. Green, E.L. (ed.) (1968) *Biology of the Laboratory Mouse*, Dover Publications Inc., New York
9. Tarkowski, A.K. (1959) *Nature*, **184**, 1286–7
10. Johnson, M.H. and Ziomek, C.A. (1981) *Cell*, **24**, 71–80
11. Ziomek, C.A., Johnson, M.H. and Handyside, A.H. (1982) *J. Exp. Zool.*, **221**, 345–55

12. Tarkowski, A.K. (1961) *Nature*, **190**, 857–60
13. Gordon, J.W., Scangos, G.A., Plotkin, D.J., Barbosa, J.A. and Ruddle, F.H. (1980) *Proc. Natl. Acad. Sci.*, **77**, 7380–4
14. Palmiter, R.D., Brinster, R.L., Hammer, R.E., Trumbauer, M.E., Rosenfeld, M.G., Birnberg, N.C. and Evans, R.M. (1982) *Nature*, **300**, 611–15
15. Lavitrano, M., Camaioni, A., Fazio, V.M., Dolci, S., Farace, M.G. and Spadafora, C. (1989) *Cell*, **57**, 717–23
16. Kollias, G., Evans, D.J., Ritter, M., Beech, J., Morris, R. and Grosveld, F. (1987) *Cell*, **51**, 21–31
17. Katsuki, M., Sato, M., Kimura, M., Yokoyama, M., Kobayashi, K. and Nomura, T. (1988) *Science*, **241**, 593–5
18. Thomas, K.R. and Capecchi, M.R. (1987) *Cell*, **51**, 503–12
19. McGrath, J. and Solter, D. (1983) *Science*, **220**, 1300–3
20. McGrath, J. and Solter, D. (1984) *Cell*, **37**, 179–83
21. Dawid, I.B. and Sargent, T.D. (1988) *Science*, **240**, 1443–8
22. Sudarwati, S. and Nieuwkoop, P.D. (1971) *Wilhelm Roux' Arch.*, **166**, 189–204
23. Rubin, G.M. (1988) *Science*, **240**, 1453–9
24. Martin, G.R. (1980) *Science*, **209**, 768–76
25. Stevens, L.C. (1970) *Dev. Biol.*, **21**, 364–82
26. Bradley, A., Evans, M., Kaufman, M.H. and Robertson, E. (1984) *Nature*, **309**, 255–6
27. Nicolas, J.F., Dubois, P., Jakob, H., Gaillard, J. and Jacob, F. (1975) *Ann. Microbiol. Inst. Pasteur*, **126A**, 1–20
28. Strickland, S. and Mahdavi, V. (1978) *Cell*, **15**, 393–403
29. Hogan, B.L.M., Taylor, A. and Adamson, E. (1981) *Nature*, **291**, 235–7
30. Edwards, M.K.S. and McBurney, M.W. (1983) *Dev. Biol.*, **98**, 187–91
31. Muramatsu, H., Hamada, H., Noguchi, S., Kamada, Y. and Muramatsu, T. (1985) *Dev. Biol.*, **110**, 284–96
32. Till, J.E. and McCulloch, E.A. (1961) *Radiation Res.*, **14**, 213–22
33. Stephenson, J.R., Axelrad, A.A., Mcleod, D.L. and Shreeve, N.M. (1971) *Proc. Natl. Acad. Sci.*, **68**, 1542–6
34. Heath, D.S., Axelrad, A.A., Mcleod, D.L. and Shreeve, M.M. (1976) *Blood*, **47**, 777–92
35. Friend, C., Scher, W., Holland, J.G. and Sato, T. (1971) *Proc. Natl. Acad. Sci.*, **68**, 378–82
36. Collins, S.J., Ruscetti, F.W., Gallagher, R.E. and Gallo, R.C. (1978) *Proc. Natl. Acad. Sci.*, **75**, 2458–62
37. Rovera, G., Santoli, D. and Damsky, C. (1979) *Proc. Natl. Acad. Sci.*, **76**, 2779–83
38. Gurdon, J.B. (1985) *Cold Spring Harbor Sym. Quant, Biol.*, **50**, 1–10
39. Tsujimoto, Y. and Suzuki, Y. (1984) *Proc. Natl. Acad. Sci.*, **81**, 1644–8
40. Jacob, F. and Monod, J. (1961) *J. Mol. Biol.*, **3**, 318–56
41. Tonegawa, S. (1985) *Sci. Amer.*, **253**(4), 104–13
42. Brown, D.D. and Dawid, I.B. (1968) *Science*, **160**, 272–80
43. Razin, A. and Riggs, A.D. (1980) *Science*, **210**, 604–10

2 Molecular architecture of the cell surface

All cells are separated from the environment by plasma membranes. The fundamental structure of the plasma membrane is a bilayer of lipids, which is about 5 nm thick [1] (Figure 2.1). Water-soluble molecules cannot move freely across the membrane; and by virtue of this plasma membrane, the interior of cells can have a chemical composition which is different from that of the environment. However, the plasma membrane is not just a steric barrier. Instead, it receives signals from outside the cells and transfers the signal into the cells. The key molecules involved in this signal transduction process are proteins anchored to the bilayer. Membrane proteins also carry out other important functions, including the uptake of specific molecules and cell adhesion. The aim of this chapter is therefore to provide basic information on the molecular architecture of the cell surface of animal cells. First there will be a brief description of the chemistry of membrane lipids and membrane proteins. Another important component of plasma membranes, carbohydrates, will be dealt with in Chapter 6. Then two important functions of plasma membrane proteins will be described – signal transduction and ion transport. Macromolecules located external to the plasma membrane, namely components of extracellular matrix will be also mentioned, together with their role in organizing extracellular structures.

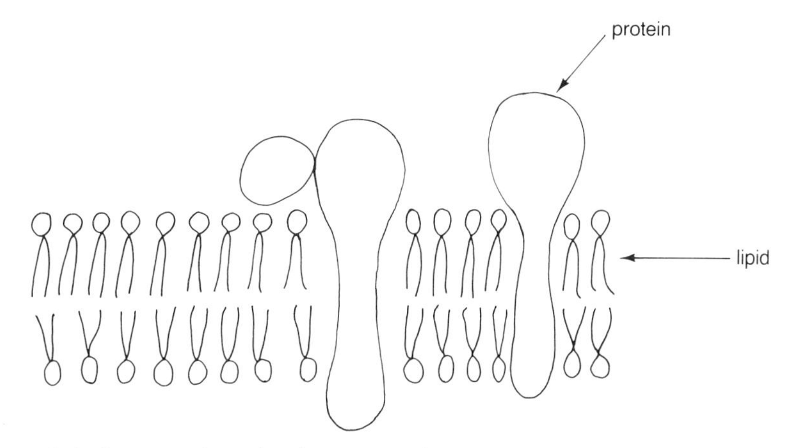

Figure 2.1 Cross section of a plasma membrane.

2.1 LIPIDS

Lipids constitute about 50% of the mass of plasma membranes in most animal cells. The most abundant membrane lipids are phospholipids such as phosphatidyl choline, phosphatidyl serine, phosphatidyl ethanolamine, phosphatidyl inositol and sphingomyelin [2] (Figure 2.2). These are amphipathic molecules having both hydrophilic (polar) and hydrophobic (non-polar) portions. The phospholic group together with a polar residue (choline, ethanolamine, serine or inositol) constitute the polar head group, while two hydrocarbon chains form non-polar tails. Except for sphingomyelin, the two hydrocarbon chains are provided by fatty acids (Figure 2.2).

The amphipathic nature is fundamental to the formation of a lipid bilayer. The non-polar groups are arranged side by side and form a sheet, while polar groups are arranged so that they face the aqueous environment. The combined result is two layers of lipids. In addition to the phospholipids, glycolipids (Chapter 6) and cholesterol constitute a significant portion of membrane lipids. They are also amphipathic molecules, and participate in the formation of the lipid bilayer.

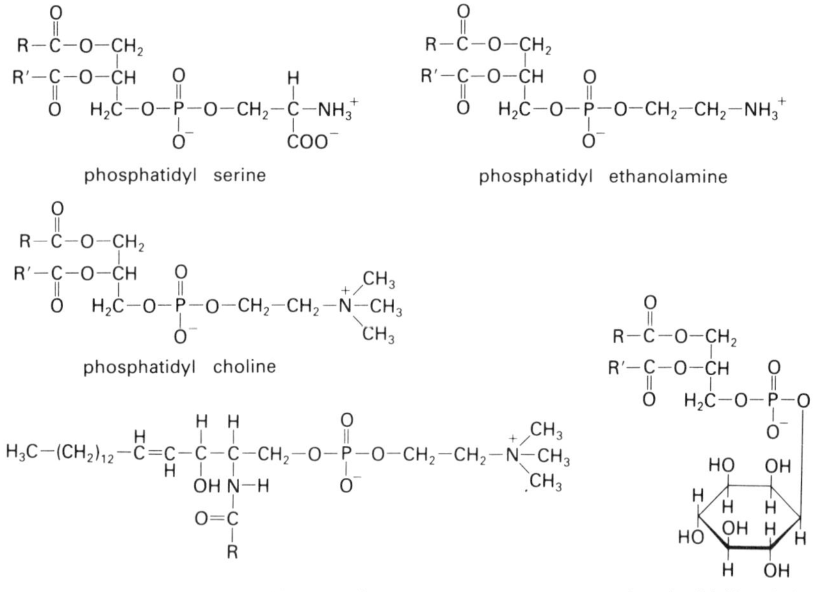

phosphatidyl serine

phosphatidyl ethanolamine

phosphatidyl choline

sphingomyelin

phosphatidyl inositol

Figure 2.2 Structures of phospholipids.

Since lipid molecules in the bilayer move rapidly by lateral diffusion, the bilayer can be considered to be two-dimensional fluid. The degree of fluidity is influenced by its composition. In particular, unsaturated fatty acids in the non-polar tail increase the fluidity by inhibiting rigid side-by-side arrangements of non-polar tails.

In spite of great lateral diffusional freedom, a flip-flop (exchange of lipid molecules in the outer layer of the bilayer with those in the inner layer) occurs only on rare occasions. Instead, lipids are asymmetrically distributed in the bilayer. Notably, all glycolipids are present in the outer half. In erythrocytes, most of phosphatidyl choline is in the outer half, while phosphatidyl ethanolamine and phosphatidyl serine are mostly found in the inner half [3].

2.2 MEMBRANE PROTEINS

Proteins constitute about 50% or less of the plasma membrane. Many of these proteins extend across the lipid bilayer [3] and are therefore called transmembrane proteins. They are ideally suited as receptors of external signals and as components of channels. The non-polar regions of these proteins are in the membrane (transmembrane domain) and serve to hold the protein to the lipid bilayer, while other regions are exposed to the extracellular and cytoplasmic sides. The transmembrane domain is composed of about 20 or more amino acids, most of which are non-polar ones (Figure 2.3). Because of the abundance of hydrophobic amino acids, this region is expected to be arranged in an α-helix [2]. Twenty amino acids in an α-helix are calculated to be 3 nm long; the figure corresponds well to the hydrophobic portion of the lipid bilayer. In most cases, N-terminal sequences are located in the external surface, while C-terminal sequences are on the cytoplasmic side. A transmembrane protein usually has only one transmembrane domain. However proteins forming channels have multiple transmembrane domains.

M Q P W L W L V F S V K L S A L W G S S A L L Q T P
S S L L V Q T N Q T A K M S C E A K T F P K G T T I
Y W L R E L Q D S N K N K H F E F L A S R T S T K G
I K Y G E R V K K N M T L S F N S T L P F L K I M D
V K P E D S G F Y F C A M V G S P M V V F G T G T K
L T V V D V L P T T A P T K K T T L K K K Q C P T P
H P K T Q K G L T C G L I T L S L L V A C I L V L L
V S L S V A I H F H C M R R R A R I H F M K Q F H K

Figure 2.3 Amino acid sequence of CD8 protein deduced from cDNA sequence (Chapter 3, [47]). The sequence starts at M (N-terminal) and ends at K (C-terminal). —, signal sequence; ~, transmembrane domain.

Plasma membrane proteins other than transmembrane proteins are bound to the lipid bilayer by interaction with other proteins (usually transmembrane proteins) or by covalent linkages to glycosyl-phosphatidyl inositol [4–6]. The structure of glycosyl-phosphatidyl inositol, when serving as the anchor for membrane proteins, has been established in a coat protein of *Trypanosoma brucei* ('variant surface glycoprotein') [4] and in Thy-1 glycoprotein [5] (Chapter 3) (Figure 2.4). Proteins with this type of anchor can be released from the membrane by a type of phospholipase C, specific either for the phosphatidyl inositol moiety or for the glycosyl-phosphatidyl inositol structure [6]. Using the enzyme as the diagnostic reagent, an

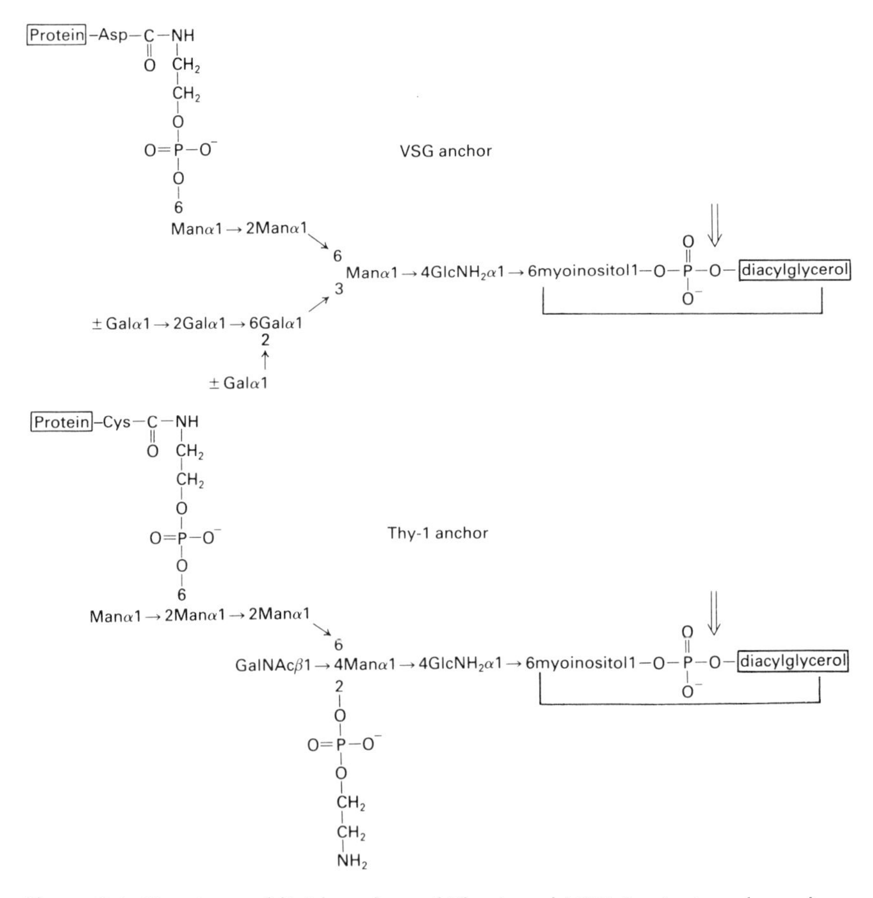

Figure 2.4 Structure of lipid anchor of Thy-1 and VSG (variant surface glycoprotein). — shows phosphatidyl inositol moiety; Arrow indicates the point of cleavage by a phospholipase C. (Based on [4]).

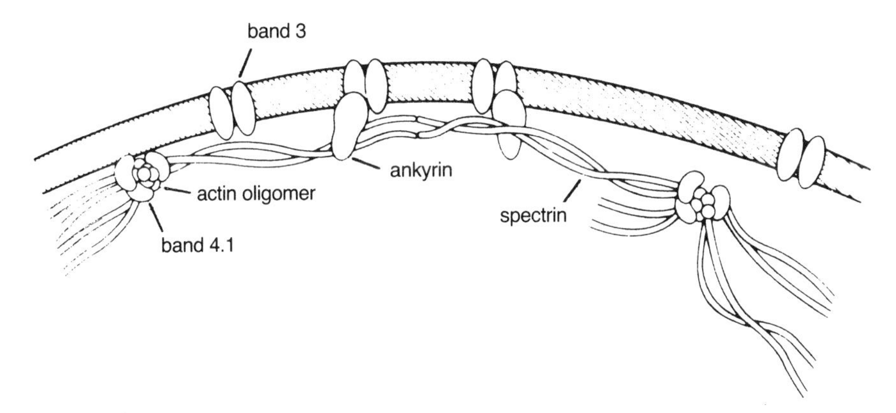

Figure 2.5 Cytoskeletal organization of erythrocytes involving band 3 (reprinted from [7]). © Cell Press.

increasing number of plasma membrane proteins have been found to have this structure.

Some transmembrane proteins exist as a monomer on the membrane plane; others form oligomers with the same molecule (homo-oligomers) or with other membrane proteins (hetero-oligomers). Many membrane proteins also form supramolecular complexes with non-membrane proteins. On the external surface, such complexes are formed betwen components of the extracellular matrix. On the cytoplasmic side, the formation of molecular complexes leads to an association of transmembrane proteins with cytoskeleton. These supramolecular complexes are believed to be important in the regulation of cellular activity by environmental information. The association of transmembrane proteins with cytoplasmic components is best illustrated in erythrocytes [7] (Figure 2.5). A transmembrane protein called band 3 forms a dimer and serves as an anion transporter in these cells. Ankyrin is attached to band 3 on the cytoplasmic side of the membrane; this protein serves as the base for holding a dimer of a filamentous protein, spectrin. Spectrin forms a filamentous network on the cytoplasmic surface of the erythrocyte membrane. Together with actin, which is also linked to band 3, the network structure is believed to be involved in the maintenance of the biconcave shape of these cells.

Since the lipid bilayer to which membrane proteins are attached is a two-dimensional fluid, many membrane proteins migrate by lateral diffusion. These proteins may be imagined as floating in a lipid bilayer as icebergs do in the sea. Indeed, some of the initial evidence for membrane fluidity came from observations on the mobility of certain transmembrane proteins at the cell surface. The 'capping' phenomenon of membrane-bound

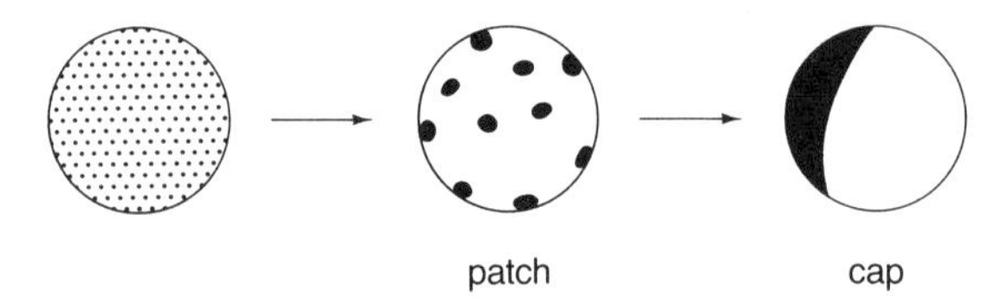

Figure 2.6 Cap formation of cell surface immunoglobulin (Ig) in the presence of antibodies to Ig. Black area represents antigen-positive region.

immunoglobulin (Chapter 3) is a typical example [8]. Immunoglobulins are distributed diffusely on the surface of lymphocytes. When antibodies against the protein are added to the cell, antibodies gradually cross link the protein floating in the plasma membrane. Thus, the diffuse location of the immunoglobulin changes to that of patchy clusters, and then all the immunoglobulin molecules become clustered at one region of the membrane to form a cap (Figure 2.6). The association of transmembrane proteins with the cytoskeletons or extracellular matrix greatly reduces their mobility. Therefore, examination of the mobility of a transmembrane protein using antibodies or other cross-linking reagents provides insight into the nature of the association.

Since transmembrane proteins are tightly integrated into the plasma membrane, they must be solubilized from the membrane, to enable the handling of the molecules by biochemical methods. Two kinds of approaches have been used frequently for this purpose – limited proteolysis and detergent treatment. Limited proteolysis using trypsin, papain or other proteases can release water soluble fragments, which can then be fractionated or analysed by usual biochemical techniques. Therefore, the method has been helpful in the study of histocompatibility antigens (Chapter 3), cell adhesion molecules (Chapter 5), etc. Detergents can solubilize intact forms of transmembrane proteins, since they destroy the lipid bilayer. The detergent-solubilized molecules interact with detergents and in many cases also with other proteins. Therefore, only selected methods are applicable for the purification of detergent-solubilized proteins. Two kinds of detergents, ionic and non-ionic, are used for the solubilization.

Sodium dodecyl sulphate (SDS) (Figure 2.7) is a typical ionic detergent. The solubilization capacity of SDS is strong, and proteins are usually denatured in its presence. When proteins solubilized by SDS are analysed by gel electrophoresis in the presence of SDS, proteins are separated according to size [9] (Figure 2.8). Thus, SDS gel electrophoresis is useful in analysing protein species in a membrane fraction. It is also helpful in purifying a

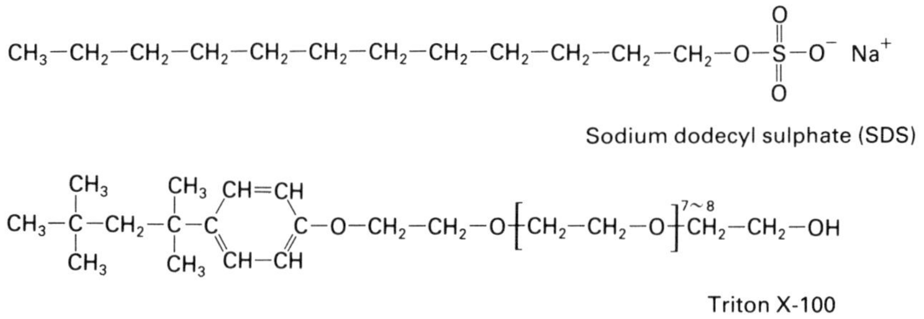

$$CH_3-CH_2-CH_2-CH_2-CH_2-CH_2-CH_2-CH_2-CH_2-CH_2-CH_2-CH_2-O-\underset{\underset{O}{\|}}{\overset{\overset{O}{\|}}{S}}-O^- \; Na^+$$

Sodium dodecyl sulphate (SDS)

$$CH_3-\underset{CH_3}{\overset{CH_3}{\underset{|}{\overset{|}{C}}}}-CH_2-\underset{CH_3}{\overset{CH_3}{\underset{|}{\overset{|}{C}}}}-C_6H_4-O-CH_2-CH_2-O\left[CH_2-CH_2-O\right]_{7\sim 8}CH_2-CH_2-OH$$

Triton X-100

Figure 2.7 Structure of SDS and Triton X-100.

membrane protein according to size; a protein band detected after electrophoresis can be excised and the protein may be eluted from the gel.

Triton X-100 (Figure 2.7) and NP-40 are non-ionic detergents. Proteins solubilized by them usually retain activity. Furthermore, ligand–protein interaction and antigen–antibody reaction can occur in the presence of non-ionic detergents. Therefore membrane proteins solubilized by these non-ionic detergents can be purified by affinity chromatography using the above interactions.

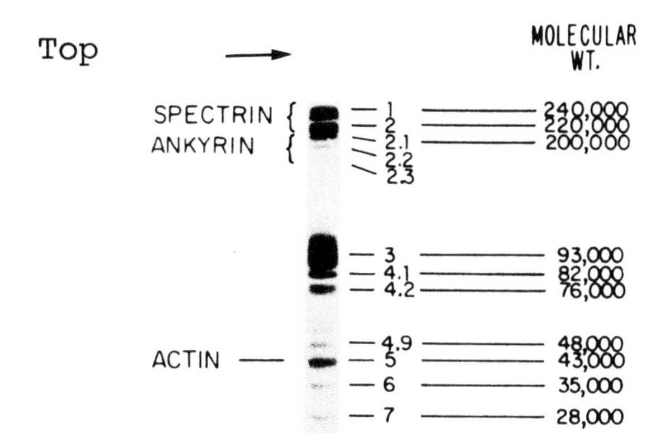

Figure 2.8 SDS gel electrophoresis of ghost (membrane and cytoskeleton fraction) of erythrocytes. Figures on the right indicate band numbers and their molecular weights. (Reprinted from [7].) © Cell Press.

Plasma membrane proteins can be isolated only in small amounts. Therefore, in the 1970s the primary structure was known for only a few membrane proteins. The introduction of the recombinant DNA technique greatly accelerated their elucidation. Knowing a partial protein sequence or having antibodies against the protein, one can clone the complementary DNA (cDNA) of the protein. From the nucleotide sequence of the cDNA clone, one can predict the sequence of the protein or its precursor.

The cloning of cDNA specifying membrane proteins is also important in elucidating the structure–function relationship of the protein. When the isolated cDNA for a plasma membrane protein is conjugated with an active promoter and introduced into a cell, the protein should be expressed in the plasma membrane. Using a cell initially lacking the protein, one can reveal or confirm the function of the protein. By introducing a mutated gene, the detailed features of the structure–function relationship can be elucidated. By introducing the gene or its modified form into fertilized eggs, it is possible to gain certain information about the function of the protein during embryogenesis and in adult tissues.

The sequence of a membrane protein predicted from the cDNA sequence is usually more than the one observed in the mature protein. Shortly after the initiator methionine, there is a stretch of 15–25 amino acids rich in hydrophobic ones (Figure 2.3). This stretch, which is called a signal sequence or leader sequence [10], is a signal for a protein being synthesized on ribosomes to attach to endoplasmic reticulum and enter its lumen. Most transmembrane proteins are synthesized by this route and they have the signal sequence. The signal sequence and occasionally some other sequences are removed by controlled proteolysis during maturation of the protein.

Post-transcriptional modifications other than proteolytic cleavage are also important. A cysteine residue can exist as cysteine itself or as cystine by forming an –S–S– bridge with other cystine residues in the peptide chain or in the neighbouring chain. An asparagine residue in an Asn–X–Thr(Ser) sequence (where X is any amino acid except proline) in the external surface is a candidate to be glycosylated [11] (Chapter 6). A serine or threonine residue in the external surface is also sometimes glycosylated. In the cytoplasmic region, a serine or threonine residue and a tyrosine residue are candidates to be phosphorylated in the hydroxyl groups. Phosphorylation of these residues may change certain functions of the proteins.

In this monograph, protein sequences are shown using one-letter symbols for amino acids (Table 2.1). Molecular weights of proteins are shown by kilodaltons in parentheses (70 K means a molecular weight 70 000). When protein sequences are compared using a computer, homologous sequences are often found in a number of proteins [12]. When two proteins are closely related, they are regarded as being in the same family. When two proteins

Table 2.1 Abbreviations for amino acids

Amino acid	Three-letter abbreviation	One-letter symbol
Alanine	Ala	A
Arginine	Arg	R
Asparagine	Asn	N
Aspartic acid	Asp	D
Cysteine	Cys	C
Glutamine	Gln	Q
Glutamic acid	Glu	E
Glycine	Gly	G
Histidine	His	H
Isoleucine	Ile	I
Leucine	Leu	L
Lysine	Lys	K
Methionine	Met	M
Phenylalanine	Phe	F
Proline	Pro	P
Serine	Ser	S
Threonine	Thr	T
Tryptophan	Trp	W
Tyrosine	Tyr	Y
Valine	Val	V

are significantly related, but the homology is not extensive, they are regarded as belonging to the same superfamily. Examples of a family are the rhodopsin family, which is involved in signal transduction (section 2.3), and the cadherin family, which is involved in cell adhesion (Chapter 5). A well-known example of a superfamily is the immunoglobulin superfamily (Chapter 3). Members of a family or a superfamily are believed to have evolved from a common ancestral gene.

2.3 SIGNAL TRANSDUCTION

As an important means of cell to cell communication, many kinds of molecules secreted by a cell affect the activities of other cells. Effector molecules have been classified into hormones, neurotransmitters and local chemical mediators [1], but the distinction between the three categories can be blurred. Among effector molecules (Figure 2.9), hydrophobic ones such as steroid hormones, thyroid hormone and retinoic acid are soluble in the lipid bilayer, and are able to enter cells without specific cell surface receptors, although they are carried by binding proteins outside the cells and by receptors inside the cells. In the nucleus the receptor–ligand complex

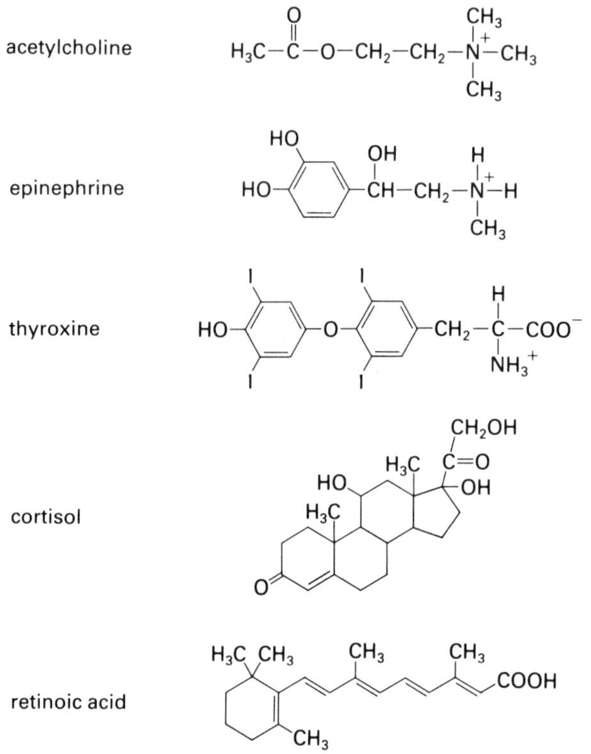

acetylcholine

epinephrine

thyroxine

cortisol

retinoic acid

Figure 2.9 Structure of low molecular weight effector molecules.

combines with a specific region of the DNA and directly affects gene expression [1].

The mode of action of hydrophilic effector molecules is more complex. Generally, they bind to cell surface receptors, which are transmembrane proteins; the signal is transduced at the cell-surface and delivered into the cells (Table 2.2). Three types of cell-surface receptors involved in signal transduction are explained below.

2.3.1 Rhodopsin family

(*a*) *Rhodopsin and G proteins*

When a receptor belongs to the rhodopsin family, the signal transduction system consists of a receptor, a GTP-binding protein (G protein) and an enzyme.

Table 2.2 Examples of signal transduction systems

External signal	Receptor	Molecules coupled with the receptor	Intracellular message
Epinephrine	β_2-adrenergic receptor	Gs, Adenylate cyclase	Cyclic AMP $\uparrow$
Light	Rhodopsin + retinol	Gt, cyclic GMP phosphodiesterase	Cyclic GMP $\downarrow$
Acetylcholine	Muscarnic acetylcholine receptor	Several G proteins coupled with enzymes and channels	Many
Acetylcholine	Nicotinic acetylcholine receptor	No	$Na^+ \uparrow$, $K^+ \downarrow$
Epidermal growth factor	Epidermal growth factor receptor	Phospholipase C? Tyrosine kinase	Inositol triphosphate $\uparrow$? Diacylglycerol $\uparrow$?

The prototype of this signal transduction system can be found in rhodopsin, a pigment present in the rod outer segment of the retina, which plays the central role in converting the light signal into a chemical one [13]. Rhodopsin (40 K) has seven transmembrane domains. When light energy is received by retinol, which is present in the cavity of rhodopsin, a change in the conformation of the rhodopsin molecule is caused. The conformational change is transmitted to a G protein, Gt (transducin), and this activates cyclic GMP phosphodiesterase. The changed level of the enzymatic activity results in a decrease of cyclic GMP, which in turn closes cation-specific channels of the retina cell, and causes hyperpolarization of the membrane. The electric signal thus formed is transmitted to dipole cells, and then to sensory neurons.

In addition to Gt, many G proteins are now known [14]. All G proteins are composed of three subunits: α (39–54 K), β (35–36 K) and γ (8–10 K). With the aid of receptors activated by the signal, G protein with bound GDP changes its conformation so that it binds to GTP rather than GDP. G protein with GTP is in an active state and activates (or inactivates) an effector molecule. Simultaneously, GTP is hydrolyzed to GDP and G protein in an inactive state is regenerated.

(*b*) β_2-*Adrenergic receptor*

A member of the rhodopsin family is also involved in the action of epinephrine, which is a hormone secreted by the adrenal medulla. One function of epinephrine is to elevate blood glucose level. This is achieved by the action of the hormone in liver cells (Figure 2.10). Cell-surface receptors for catecholamines including epinephrine are called adrenergic receptors, and are classified into α_1, α_2, β_1 and β_2 based on their pharmacological properties. The liver receptor is a β_2-adrenergic receptor. The receptor from the lung (64 K) was solubilized by a detergent, digitonin, and purified by affinity chromatography using a ligand, alprenolol, followed by high performance liquid chromatography. A partial amino acid sequence from the isolated receptor was utilized to clone the cDNA of the receptor. The predicted sequence of the receptor is homologous to the sequence of rhodopsin; notably, the seven transmembrane segments are also present in the receptor [15, 16] (Figure 2.11). The signal received by the β_2-adrenergic receptor is transferred to Gs (stimulatory G protein), which in turn activates adenylate cyclase. The enzyme forms 3′, 5′-cyclic AMP from ATP, and cyclic AMP activates a protein kinase (cyclic AMP-dependent protein kinase; A kinase) which phosphorylates the hydroxyl groups of serine and threonine in many proteins. The protein kinase eventually activates the glycogen-degrading enzyme (phosphorylase) and inactivates glycogen synthetase. The stimulated glycogen degradation yields much free glucose,

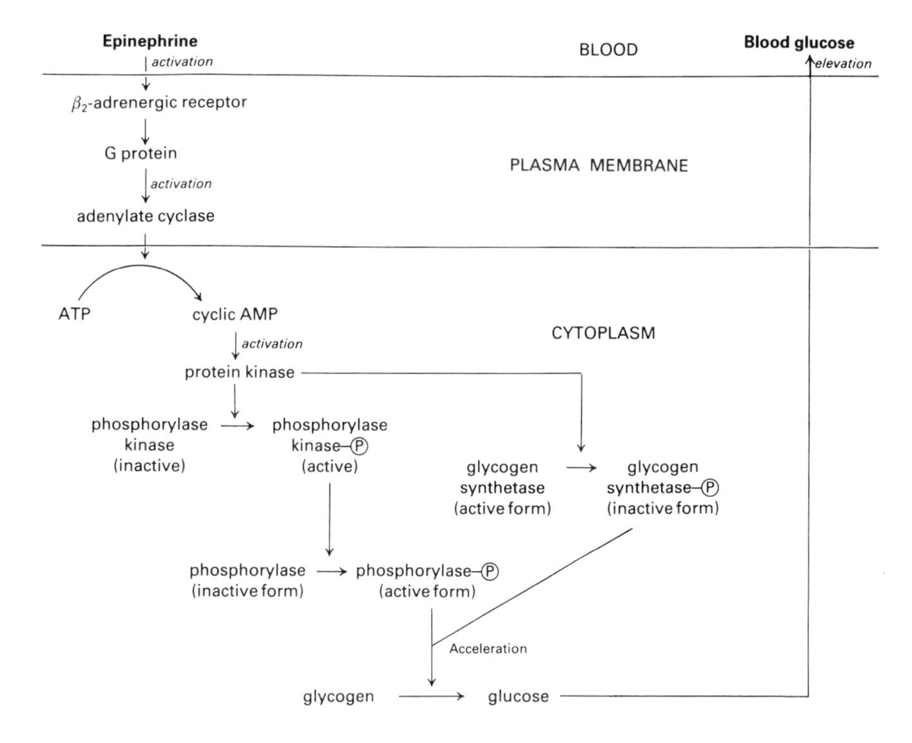

Figure 2.10 Mechanism by which epinephrine raises blood glucose levels.

which is used to elevate the blood glucose level [2] (Figure 2.10). On the other hand, α_2-adrenergic receptor, which also belongs to the rhodopsin family [17], couples with Gi (inhibitory G protein), which inhibits adenylate cyclase. Therefore, epinephrine can decrease cyclic AMP level in cells with the α_2-receptor: by changing the receptor and G protein, the same ligand can cause opposite effects in target cells.

(*c*) *Muscarinic acetylcholine receptor*

The muscarinic acetylcholine receptor is another important member of the rhodopsin family. Acetylcholine is employed as the neurotransmitter of the cholinergic neuron. Acetylcholine receptors are classified into nicotinic and muscarinic receptors based on pharmacological properties. The former is found in the neuromuscular junction and the latter in many tissues, including neural tissues. The primary structure of the muscarinic

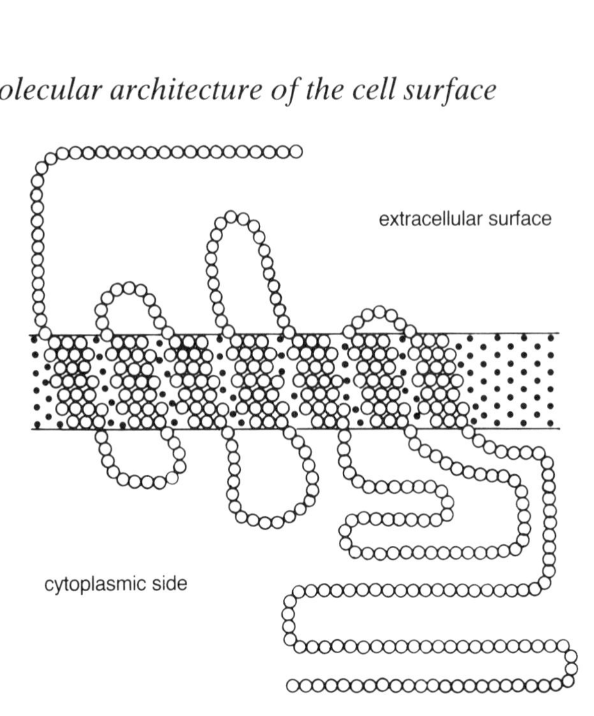

Figure 2.11 β_2-adrenergic receptor spans the membvrane 7 times. A circle represents an amino acid. (Modified from [16].)

acetylcholine receptor is homologous to rhodopsin and the β_2-adrenergic receptor [18]. The activation of the muscarinic acetylcholine receptor elicits different reactions by coupling with different G proteins; the main reactions are modulation of the K^+ channel, inhibition of adenylate cyclase and activation of a specific phospholipase C (phosphoinositidase) which cleaves phosphatidyl inositol 4,5-bisphosphate (Figure 2.12).

(*d*) *Phospholipase C, inositol 1,4,5-triphosphate and protein kinase C*

Many factors other than acetylcholine, such as platelet-derived growth factor, fibroblast growth factor, vasopressin and thrombin, activate a phospholipase C (phosphoinositidase) through respective receptors and G proteins [19, 20]. The two products released from phosphatidyl inositol 4,5-bisphosphate by the action of the phospholipase C are important in the regulation of many cellular activities. One product, inositol 1,4,5-triphosphate works in the cytoplasm, while the other, diacylglycerol, functions in the membrane plane [20] (Figure 2.12). The inositol phosphate triggers the release of Ca^{2+} from endoplasmic reticulum. The released ion exerts many effects, mainly by activating proteins which require a certain concentration of Ca^{2+} for their action. For example, Ca^{2+} activates calmodulin, which then activates a protein kinase (calmodulin-dependent

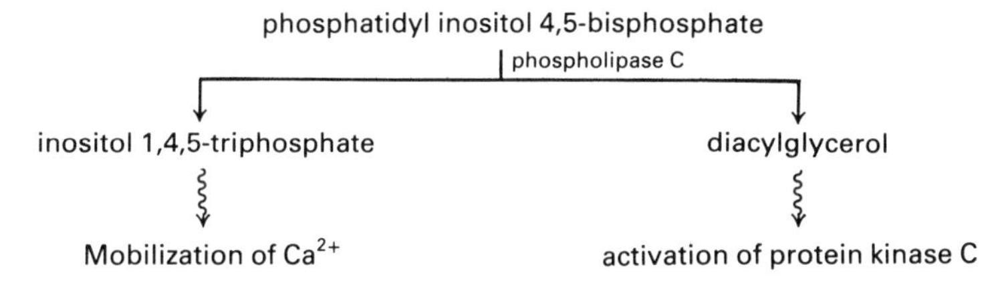

Figure 2.12 Two products released by phospholipase C.

kinase). On the other hand, diacylglycerol activates protein kinase C (C kinase), which comprises a group of protein kinases phosphorylating serine and threonine residues in proteins and requiring diacylglycerol, phosphatidyl serine and Ca^{2+} for their action [21, 22, 23]. Phorbol esters (TPA), a class of tumour promoter, mimic diacylglycerol and activate C kinase. The protein kinase is believed to be involved in the regulation of a number of cellular activities, since the tumour promoter and specific inhibitors of the protein kinase exert many physiological effects. Furthermore, when isolated cDNA of C kinase was introduced in an expression vector into rat fibroblasts, they lost their normal growth control and became cancerous [24]. One of the roles of C kinase is probably feedback inhibition of the inositolphosphate/diacylglycerol system by phosphorylating the receptor and by stimulating the removal of Ca^{2+} from the cytoplasmic component [20].

2.3.2 Nicotinic acetylcholine receptor

The nicotinic acetylcholine receptor is present in the neuromuscular junction. Receiving acetylcholine released from the nerve terminal, this receptor, present in the membrane of the muscle cells, causes a rapid increase in permeability to both Na^+ and K^+; this leads to depolarization of the membrane of the acceptor cells. Nicotinic acetylcholine receptor was isolated from electric eel by solubilization with a detergent luburol, affinity chromatography using a ligand, bromoacetylcholine bromide, and further purification by gel filtration in the presence of a detergent. The receptor is composed of 4 kinds of subunits – α (40 K), β (50 K), γ (60 K) and δ (65 K) in a ratio of 2 : 1 : 1 : 1. These subunits are arranged to form a channel in the plasma membrane [25]. The binding of acetylcholine to the external side of the receptor causes an opening of the channel. Based on the partial amino acid sequence, cDNA clones were isolated for all the subunits and their primary structures were elucidated. They have sequence homology to each other, and at least 4 transmembrane domains are evident for each subunit [26, 27, 28]. The transmembrane domains are involved in the formation of the channel. Nicotinic acetylcholine receptor is the simplest form of the signal transduction system so far elucidated. It receives the signal, and performs the effector function in the same molecule.

2.3.3 Receptors with a tyrosine kinase domain

Epidermal growth factor (EGF) is a peptide factor which promotes the growth of epidermal cells (Chapter 4). The receptor for EGF is a transmembrane protein of molecular weight 170 K. The cytoplasmic side was found to have the activity of a protein kinase, which phosphorylates tyrosine residues in proteins [29] (Figure 2.13). The receptors for platelet-derived growth factor (PDGF; Chapter 4) and colony stimulation factor-1 (CSF-1; Chapter 4) also possess the cytoplasmic tyrosine kinase domain [30, 31]. In these cases 2 instead of 1 kinase domains are present (Figure 2.13).

Insulin is a typical example of a hormone; it enhances the uptake of glucose into many cells and also promotes cellular growth. The insulin

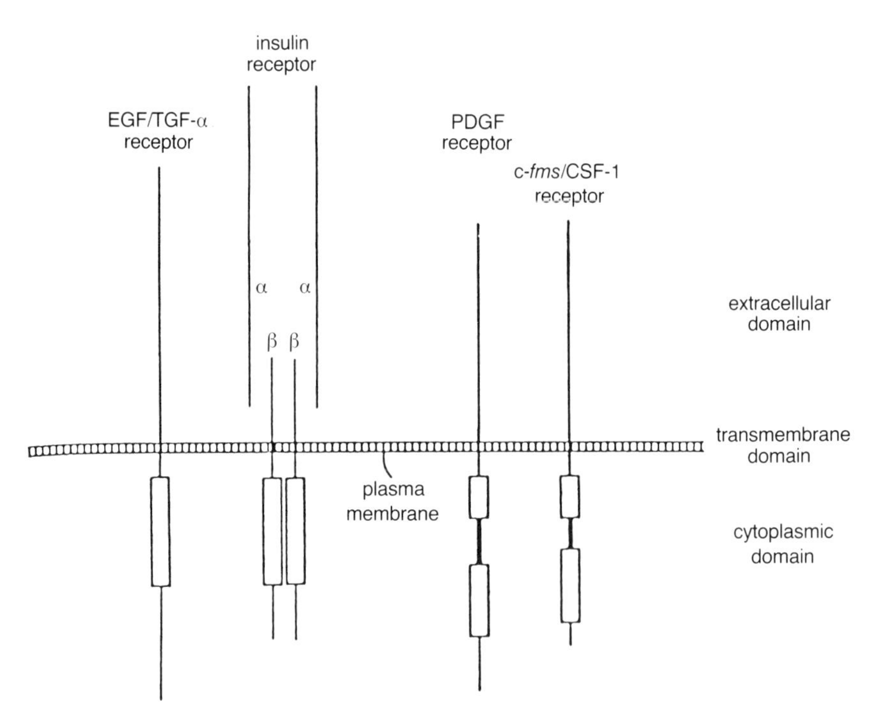

Figure 2.13 Organization of receptors for some growth factors. □ shows a tyrosine kinase domain. (Reprinted from [32].) © 1985. Macmillan Magazines Ltd.

receptor is composed of 4 subunits – 2α (120–130 K) and 2β (90 K). The β-subunit is the transmembrane polypeptide, and again possesses the tyrosine kinase domain [32, 33] (Figure 2.13). As in all cases of tyrosine kinase-type receptors, the cytoplasmic tails are long (500–600 amino acid residues), and the receptors cross the membrane only once; this structural feature permits the distinction of the tyrosine kinase-type receptors from other classes of receptors [34].

Tyrosine kinase activity was originally found in a product of the oncogene of avian sarcoma virus (*src*) [35]. Subsequently, several other oncogenes were also found to specify tyrosine kinases. Furthermore, sequence homologies were revealed between the product of the oncogenes and the kinase domain of the receptors; namely *erb B* (an oncogene of avian erythroblastosis virus) and EGF receptor [29], *fms* (an oncogene of McDonough strain of feline sarcoma virus) and CSF-1 receptor [30], *ros* (an oncogene of avian sarcoma virus UR2) and insulin receptor [32]. These findings gave rise to the belief that the tyrosine kinase domain of the receptors plays a central role in signal transduction. Indeed, mutated receptors which lack the tyrosine kinase domain are unable to transduce signals [34]. The physiological substrate of EGF receptor tyrosine kinase appears to be a phospholipase C triggering phosphatidyl inositol 4,5-bisphosphate metabolism [36, 37]. Thus, the signal transduction system involving inositol 1,4,5-triphosphate and diacylglycerol may be activated by two distinct routes, G protein (section 2.3.1 (d)) and tyrosine kinases.

Insulin is also known to activate a specific phospholipase C which degrades glycosyl-phosphatidyl inositol. One of the released products, inositolphosphate glycan, has been proposed to be the major mediator of insulin action. The tyrosine kinase domain of insulin receptor may be important in the activation of the phospholipase C.

2.3.4 Comments

Typical examples of signal transduction have been reviewed. It is important to remember that in a cell, multiple signal transduction systems interact with each other [20]. Furthermore, multiple signal transduction systems may work simultaneously to regulate a cellular activity. For example, glycogen synthetase is inhibited by phosphorylation both by A kinase and C kinase.

Signal transduction must be very important in the regulation of differentiation. Thus, important components of signal transduction systems, such as transmembrane receptors, G proteins, protein kinases, phospholipase C, inositolphosphate, cyclic AMP and Ca^{2+} are important clues in analysing the mechanism of differentiation.

The primary structures of many protein kinases have been elucidated [38]. There are amino acid residues conserved in all protein kinases.

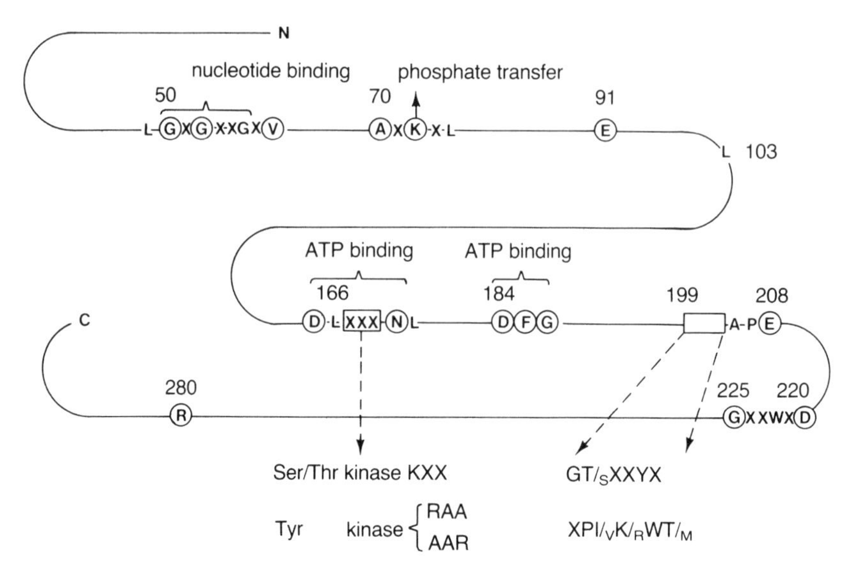

Figure 2.14 Consensus sequences of protein kinases. Figures indicate amino acid numbers in the α-subunit of a cyclic AMP-dependent protein kinase. Encircled amino acids are those conserved in most cases. X is any amino acid. The mode of protein folding is arbitrary. (Data taken from [38].)

Furthermore, some amino acid sequences are specific for serine/threonine kinases, while others are specific for tyrosine kinases (Figure 2.14). This information is helpful when one examines whether a protein sequence has a kinase domain.

We should remember that all signal transduction systems may not fall into the 3 types so far described. Indeed, sequences of nerve growth factor (Chapter 4) receptor [39, 40] and of IL-2 (Chapter 4) receptor [41] do not have homologies to the 3 types of receptors. Furthermore, the most critical question, which is largely unanswered, is how cytoplasmic messages produced by signal transduction are delivered to the nucleus and regulate gene expression (Chapter 7).

2.4 CHANNEL PROTEINS

A special device must be present in plasma membranes in order to allow a polar molecule to pass. The mode of membrane transport differs according to the nature of a polar molecule; ions use channels, while high molecular weight materials are endocytosed. For endocytosis, a high molecular weight substance binds with a specific receptor at the cell surface (low density lipoprotein receptor [42], transferrin receptor [43], asialoglycoprotein receptor [44], etc.), and the ligand-receptor complex is internalized with the aid of coated pits that are present on the cytoplasmic side of the complex. Clathrin is the major component of coated-pits and this has a three-legged structure composed of 3 heavy chains (192 K) and 3 light chains (23–25 K) [45].

The molecular nature of several channel proteins have been clarified recently. Since the ionic status of a cell must be important in differentiation, channel proteins will be described briefly below. It should be mentioned that this area is one of the most advanced in the study of the molecular biology of plasma membranes.

Channels can be classified into two types. When the transport is against the concentration gradient across the membrane (active transport), it consumes the free energy of ATP, and the channel molecule has the character of an ATPase. Na^+, K^+ ATPase is a typical example. When the transport is not against the concentration gradient (passive transport), the free energy of ATP is not required, and the ATPase structure is absent in the channel molecule. However, a 'gate' structure is required so that a channel opens only under selected physiological conditions. The channel proteins described here (except for Na^+, K^+ ATPase), and nicotinic acetylcholine receptor mentioned in the previous section, fall into this category.

2.4.1 Na^+, K^+ ATPase

Cells actively pump out Na^+ and pump in K^+, so that the concentration of Na^+ is about 10 times lower inside the cell than it is outside and that of K^+ is about 10 times higher inside the cell. This disequilibrium is attained by Na^+, K^+ ATPase, which expels 3 moles of Na^+ and takes in 2 moles of K^+ using the free energy generated by the hydrolysis of one mole of ATP. The enzyme is specifically inhibited by ouabain, and this reagent can be used as a diagnostic tool to determine whether the ATPase is involved in a physiological process. Na^+, K^+ ATPase consists of two types of subunits, α (110 K) and β (35 K), and the primary structure of both subunits was established from the sequence of the corresponding cDNA [46, 47]. The β-subunit contains a multiple membrane spanning region, which constitutes a channel through which Na^+ and K^+ are transported.

2.4.2 Na^+ channel

In nerve cells, the disequilibrium of Na^+ and K^+ ions generated by Na^+,K^+ ATPase forms the basis of the conductance of electrical signals in nerve cells. The voltage-gated Na^+ channel is distributed over the surface of nerve plasma membrane. When the membrane potential is lowered the Na^+

```
Ca  K P F E T I I L L T I F A N C V A L A V Y L P M P E D D N N S L N L G L E K L E Y F F L T V F S I E   100
Na  S L F N V L I M C T I L T N C V F M T M - - - - - - - - - S N P P D W T - K N V E Y T F T G I Y T F E   169
                        S1                                                           S2

Ca  A A M K I I A Y G F L F H Q D A Y L R S G W N V L D F I I V F L G V F T A I L E Q V N V I Q S N T A   150
Na  S L I K I L A R G F C L E D F T F L R N P W N W L D F T V I T F A Y V T E F V N L G N - - - - - - - -   212
                                                  S3

Ca  P M S S K G A G L D V K A L R A F R V L R P L R L V S G V P S L Q V V L N S I F K A M L P L F H I A   200
Na  - - - - - - - - - - V S A L R T F R V L R A L K T I S V I P G L K T I V G A L I Q S V K K L S D V M   252
                                  S4

Ca  L L V L F M V I I Y A I I G L E L F K G K M H K T C Y Y I G T D I V A T V E N E K P S P C A R T G S   250
Na  I L T V F C L S V F A L I G L Q L F M G N L R N K C L Q W P P D N - S T F E I N I T S L C G N S S D   343
                S5

Ca  G R P C T I N G S E C R G G W P G P N H G I T H F D N F G F S M L T V Y Q C I T M E G W T D V L Y W   300
Na  A G Q C P - E G Y I C V K A G R N P N Y G Y T S F D T F S W A F L S L F R L M T Q D F W E N - L Y Q   391

Ca  V N D A I G N E W P W I Y F V T L I L L G S F F I L N L V L G V L S G                                 335
Na  L T L R A A G K T Y M I F F V L V I F L G S F Y L I N L I L A V V A M                                 426
                                S6
```

Figure 2.15 Homology between repeat 1 of Ca^{2+} channel and that of Na^{+} channel (cited from [49]).

channel opens and allows the entry of external Na^+ into the cells. The membrane potential itself is maintained by the unequal distribution of Na^+ and K^+ and the loss of membrane potential is called depolarization. Thus, when a region of the membrane is depolarized by an external signal, this depolarization is conducted by successive opening of the neighbouring Na^+ channels. The Na^+ channel from an electric eel is composed of a single polypeptide of molecular weight 260–300 K. The primary structure of the polypeptide was determined by sequencing the corresponding cDNA clone [48]. The protein exhibits four repeating homologous units. Each unit contains 6 segments, most of which are hydrophobic and constitute the transmembrane domain (Figure 2.16).

2.4.3 Ca^{2+} channel

The Ca^{2+} channel is distributed in a wide variety of cells; it serves to trigger the release of transmitters, trigger muscle contraction and so on. The primary structure of a Ca^{2+} channel protein (1,4-dihydropyridine-binding protein; 170 K) was deduced in skeletal muscle by cDNA cloning [49]. Extensive homology exists between the Ca^{2+} channel protein and the Na^+ channel protein (Figure 2.15). The Ca^{2+} channel protein is also composed of 4 repeating homologous units. Each unit contains 6 segments. Segments 1,2,3,5 and 6 are hydrophobic, segment 4 usually contains 5–6 arginine and/or lysine residues at every third position. This structural feature is also seen in Na^+ channel protein. Apparently, all the 6 segments span the membrane and contribute to form a channel (Figure 2.16). The positive charge in segment 4 may function in channel gating.

2.4.4 K^+ channel

Several types of K^+ channels are known. Among them, a voltage-sensitive type called A channel whose structure was elucidated by molecular genetic studies using the *Drosophila* mutant *Shaker*. The name *Shaker* comes from

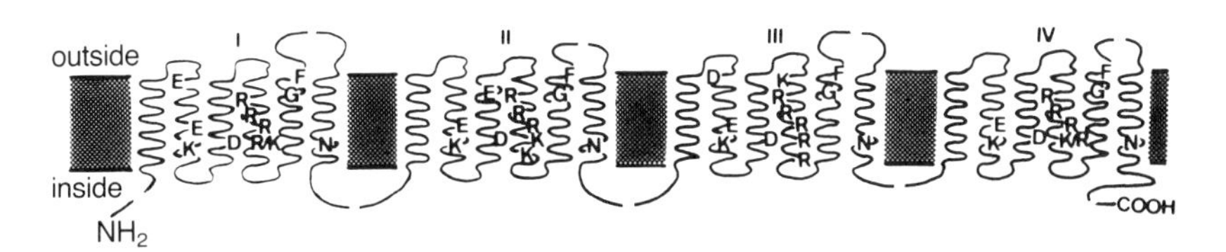

Figure 2.16 Organization of Ca^{2+} channel protein in the membrane. Each repeat (I–IV) has 6 transmembrane regions. (Cited from [49].) © 1987. Macmillan Magazines Ltd.

the fact that the mutant fly shakes its legs when anaesthetized with ether. The *Shaker* mutants are characterized by poor motor control, abnormal synaptic efficiency, and spontaneous excitability. Evidence indicates that the *Shaker* gene specifies the A channel protein. With the aid of *Drosophila* genetics, the *Shaker* gene was cloned and the structure of the channel protein was predicted [50]. The K^+ channel protein (64–74 K) is shorter than Na^+ and Ca^{2+} channel proteins. Nevertheless, the K^+ channel protein is homologous with Na^+ and Ca^{2+} channel proteins. The K^+ channel protein also has 6 segments and most amino acids conserved between Na^+ and Ca^{2+} channel proteins are also conserved in the K^+ channel protein. Several components are found in transcripts of the *Shaker* locus. Thus, alternative splicing can generate a diversity of K^+ channels.

2.4.5 Gap junction

Gap junctions are channels directly linking the adjacent cells and allowing an exchange of ions and small metabolites [51, 52]. They are present in most organs, and cells linked by gap junctions communicate with each other extensively. An elevation of Ca^{2+} and fall in pH cause the closure of the channel. The major protein constituting gap junctions appears to be the '27 K' protein. From the cDNA sequence, the protein was predicted to span the membrane 4 times [53].

2.5 EXTRACELLULAR MATRIX

The extracellular matrix of most cells is occupied by a network of macromolecules. They play important roles not only in maintaining the structure of the animal body but also in regulating various cellular activities.

2.5.1 Collagens

The most conspicuous component in the extracellular matrix of vertebrate cells is the collagen fibre, which extends up to many μm. Collagens constitute about 25% of the total body protein. The structural unit of a collagen fibre is tropocollagen, which is a long fibrous protein (300×1.5 nm) and is composed of three twisted polypeptide chains called α-chains (100 K) [54]. Tropocollagen is arranged to form a long collagen fibre (Figure 2.17). In the fibre, tropocollagen is usually cross-linked to form an insoluble structure.

Collagens have a characteristic amino acid composition; they are rich in glycine and proline and contain two unusual amino acids, namely hydroxylysine and hydroxyproline. Notably, glycine is present as almost every third amino acid [54].

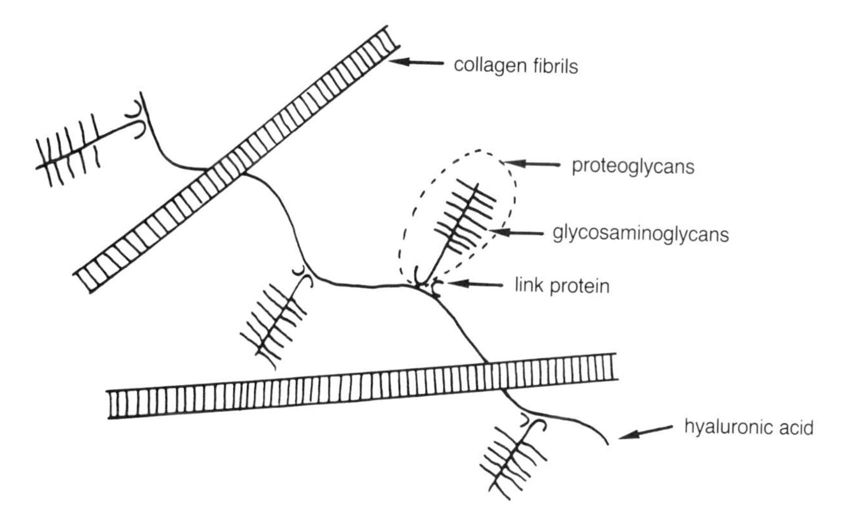

Figure 2.17 Molecular organization in extracellular matrix (based on [54]).

There are several types of collagens. The primary structure of the subunits in different types of collagen can be distinguished from each other [55]. Collagen types are distinctly distributed in different tissues; for example, type IV collagen is found almost exclusively in basement membrane.

2.5.2 Glycosaminoglycans and proteoglycans

Glycosaminoglycans are also important components of the extracellular matrix [54, 56]. They are linear polysaccharides, which are principally composed of a disaccharide unit containing *N*-acetylglucosamine or *N*-acetylgalactosamine (Chapter 6). Because of the presence of uronic acids or sulphate, or both, they are highly acidic carbohydrates. Hyaluronic acid, chondroitin sulphates, heparan sulphate, heparin and keratan sulphate are well known examples of glycosaminoglycans (Figure 2.18). Except for hyaluronic acid, glycosaminoglycans are covalently linked to proteins, and the resultant molecules are called proteoglycans. Furthermore in cartilage, hyaluronic acid is non-covalently bound to proteoglycans by a protein called link protein (Figure 2.17). The best characterized proteoglycan is that of cartilage. A typical molecule (around 2 500 K) consists of a core protein (200–300 K), about 100 chondroitin sulphate chains and up to 130 keratan sulphate chains [57]. Proteoglycans are expected to include diverse kinds of molecules [57]. Indeed, molecular biological studies are yielding interesting results on the structural diversity and function of the core proteins. For example, Ser–Gly sequences, whose serine residue serves as the acceptor

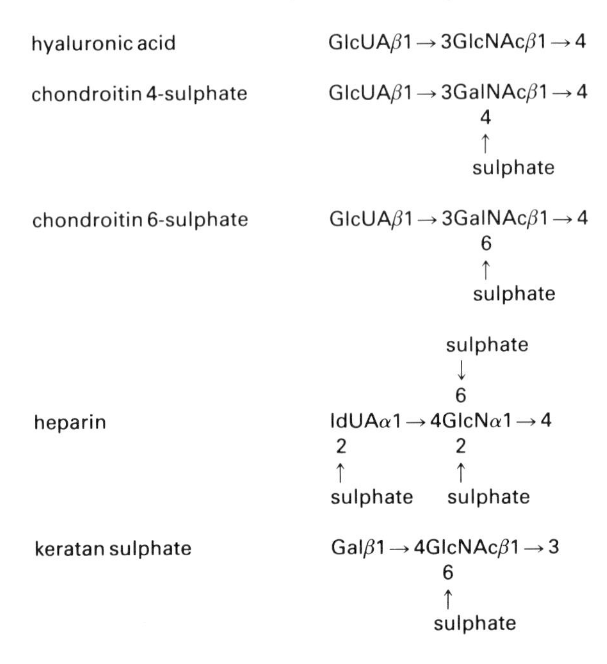

Figure 2.18 Fundamental structural unit of glycosaminoglycans. GlcUA: glucuronic acid; IdUA: iduronic acid; GlcN: glucosamine; GlcNAc: *N*-acetylglucosamine; GalNAc: *N*-acetylgalactosamine; Gal: galactose. The structures of these monosaccharides are summarized in Chapter 6.

site for glycosaminoglycan chains, are randomly repeated in a proteoglycan from rat yolk sac tumour (Figure 2.19) [58], while they are only sparsely present in proteoglycans from other sources [59, 60]. In some proteoglycans, sequences homologous to that of animal lectins were found in the C-terminal region [61, 62]. Furthermore, an epidermal growth factor-like sequence was detected in a proteoglycan from fibroblasts [62]. Thus, the core protein is expected to serve not only as the anchor of glycosaminoglycan chains but also as a functional protein interacting with other molecules.

R G F P N D F F P I S D D Y S G S G S G S G S G S G
S G S G S G S G S G S G S G S G S G S G S G S G S G
S G S G S G S G S G S L A D M E W E Y Q P T D E N N
I V Y F N Y G P F D R M L T E Q N Q E Q P G D F I I

Figure 2.19 Partial amino acid equence of a proteoglycan core protein. The sequence starts at R (N-terminal side) and ends at I (C-terminal). (Based on [58].)

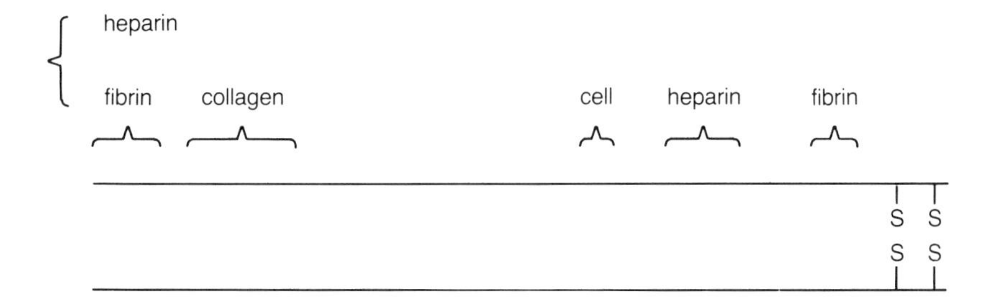

Figure 2.20 Functional domain of fibronectin polypeptides.

2.5.3 Fibronectin and laminin

Many other proteins are also detected in the extracellular matrix. Among them, fibronectin and laminin are highly important, and both of them serve to connect collagens and proteoglycans with the cell surface.

Fibronectin is composed of two subunits (220 K), which are held together by an –S–S– bridge. The protein has multiple domains, namely fibrin-binding, collagen-binding, cell-binding and heparin-binding (Figure 2.20) [63]. The cell-binding domain has an arginine–glycine–aspartic acid (RGD) sequence, which is recognized by a fibronectin receptor on cell surface [64]. Because of the multi-domain nature, fibronectin can mediate cell attachment to the collagen substratum. Fibronectin is greatly reduced in fibroblasts transformed by certain oncogenic viruses. When fibronectin was added to the culture of transformed cells, they resumed the spindle-shaped morphology and ordered arrangements of normal cells [65]. This finding was a major breakthrough in cell surface research in the 1970s.

Laminin is a basement membrane protein of molecular weight 900–800 K; it has an extended cruciform structure and appears to consist of three chains, A (400 K), B1 (230 K) and B2 (220 K) (Figure 2.21) [66, 67]. Laminin has multiple domains reacting with type IV collagen, heparin, a sulpho-glycolipid (Chapter 6) and a cell surface receptor for laminin [68]. Laminin promotes the adhesion and growth of epithelial cells on a substratum of type IV collagen [69]. Furthermore, an outgrowth of neurite is promoted by laminin [70]. Laminin B1 chain was found to have a cysteine-rich repeat typically found in epidermal growth factor [71].

Still another matrix protein appears to be of general importance for cellular function. Cytotactin (also called Tenascin or J1) is a glycoprotein composed of subunits of different molecular weights (220, 200 and 190 K) [72]. Fab′ fragment of antitenascin antibodies inhibited glia-neuron adhesion. Since the glycoprotein was selectively detected in mesenchymal

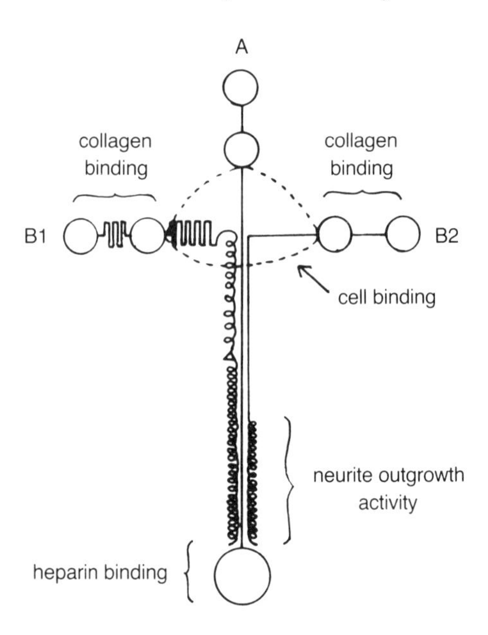

Figure 2.21 Subunit organization and domain function proposed for laminin (cited from [71]).

tissue surrounding fetal rat mammary glands, hair follicles and teeth, it was proposed to be involved in the mesenchymal–epithelial interaction [73]. Tenascin binds to fibronectin, and one of its functions may be to inhibit fibronectin action [74]. Tenascin also binds to proteoglycans, and its interaction with cell surface proteoglycans may enable the glycoprotein to be involved in cell–cell and cell–substratum adhesion.

REFERENCES

1. Alberts, B., Bray, D., Lewis, J., Raff, M., Roberts, K. and Watson, J.D. (1989) *Molecular Biology of the Cell*, Garland Publishing Co., New York
2. Stryer, L. (1988) *Biochemistry*, W.H. Freeman and Co., New York
3. Bretscher, M.S. (1973) *Science*, **181**, 622–9
4. Ferguson, M.A.J., Homans, S.W., Dwek, R.A. and Rademacher, T.W. (1988) *Science*, **239**, 753–9
5. Homans, S.W., Ferguson, M.A.J., Dwek, R.A., Rademacher, T.W., Anand, R. and Williams, A.F. (1988) *Nature*, **333**, 269–72
6. Ferguson, M.A.J. and Williams, A.F. (1988) *Ann. Rev. Biochem.*, **57**, 285–320
7. Branton, D., Cohen, C.M. and Tyler, J. (1981) *Cell*, **24**, 24–32
8. Taylor, R.B., Duffus, W.P.H., Raff, M.C. and de Petris, S. (1971) *Nature New Biol.*, **233**, 225–9
9. Shapiro, A.L., Vinuela, E. and Maizel, J.J. (1967) *Biochem. Biophys. Res. Commun.*, **28**, 815–20

10. von Heijne, G. (1984) *J. Mol. Biol.*, **173**, 243–51
11. Marshall, R.D. and Neuberger, A. (1972). In *Glycoproteins*, (ed. A. Gottschalk) Elsevier, Amsterdam, Netherlands, pp. 453–70
12. Dayhoff, M.O., Barker, W.C. and Hunt, L.T. (1983) *Methods Enzymol.*, **91**, 524–45
13. Engelman, D.M., Goldman, A. and Steiz, T.A. (1982) *Methods Enzymol.*, **88**, 81–8
14. Lochrie, M.A. and Simon, M.I. (1988) *Biochemistry*, **27**, 4957–65
15. Dixon, R.A.F., Kobilka, B.K., Strader, D.J., Benovic, J.L., Dohlman, H.G. *et al.* (1986) *Nature*, **321**, 75–9
16. Dohlman, H.G., Bouvier, M., Benovic, J.L., Caron, M.G. and Lefkowitz, R.J. (1987) *J. Biol. Chem.*, **262**, 14282–8
17. Kobilka, B.K., Matsui, H., Kobilka, T.S., Yang-Feng, T.L., Francke, U. *et al.* (1987) *Science*, **238**, 650–6
18. Kubo, T., Fukuda, K., Mikami, A., Maeda, A., Takahashi, H. *et al.* (1986) *Nature*, **323**, 411–16
19. Hokin, L.E. (1985) *Ann. Rev. Biochem.*, **54**, 205–35
20. Berridge, M.J. (1987) *Ann. Rev. Biochem.*, **56**, 159–93
21. Nishizuka, Y. (1984) *Nature*, **308**, 693–8
22. Nishizuka, Y. (1988) *Nature*, **334**, 661–5
23. Coussens, L., Parker, P.J., Rhee, L., Yang-Feng, T.L., Chen, E. *et al.* (1986) *Science*, **233**, 859–66
24. Housey, G.M., Johnson, M.D., Hsiao, W.L.W., O'Brien, C.A., Murphy, J.P. *et al.* (1988) *Cell*, **52**, 343–54
25. Heidmann, T. and Changeux, J.-P. (1978) *Ann. Rev. Biochem.*, **47**, 317–57
26. Noda, M., Takahashi, H., Tanabe, T., Toyosato, M., Furutani, Y. *et al.* (1982) *Nature*, **299**, 793–7
27. Noda, M., Takahashi, H., Tanabe, T., Toyosato, M., Kiyotani, S. *et al.* (1983) *Nature*, **301**, 251–5
28. Noda, M., Takahashi, H., Tanabe, T., Toyosato, M., Kikyotani, S. *et al.* (1983) *Nature*, **302**, 528–32
29. Downward, J., Yarden, Y., Mayes, E., Scrace, G., Totty, N. *et al.* (1984) *Nature*, **307**, 521–7
30. Sherr, C.J., Rettenmier, C.W., Sacca, R., Roussel, M.F., Look, A.T. and Stanley, E.R. (1985) *Cell*, **41**, 665–76
31. Yarden, Y., Escobedo, J.A., Kuang, W.-J., Yang-Feng, T.L., Daniel, T.O. *et al.* (1986) *Nature*, **323**, 226–32
32. Ullrich, A., Bell, J.R., Chen, E.Y., Herrera, R., Petruzzelli, L.M. *et al.* (1985) *Nature*, **313**, 756–61
33. Rosen, O.M. (1987) *Science*, **237**, 1452–8
34. Yarden, Y. and Ullrich, A. (1988) *Ann. Rev. Biochem.*, **57**, 443–78
35. Collett, M.S. and Erikson, R.L. (1978) *Proc. Natl. Acad. Sci.*, **75**, 2021–4
36. Margolis, B., Rhee, S.G., Felder, S., Mervic, M., Lyall, R. *et al.* (1989) *Cell*, **57**, 1101–7
37. Meisenhelder, J., Suh, P., Rhee, S.G. and Hunter, T. (1989) *Cell*, **57**, 1109–22
38. Hanks, S.K., Quinn, A.M. and Hunter, T. (1988) *Science*, **241**, 42–52
39. Radeke, M.J., Misko, T.P., Hsu, C., Herzenberg, L.A. and Schooter, E.M. (1987) *Nature*, **325**, 593–7
40. Johnson, D., Lanahan, A., Buck, C.R., Sehgal, A., Morgan, C. *et al.* (1986) *Cell*, **47**, 545–54
41. Waldmann, T.A. (1986), *Science*, **232**, 727–32

42. Yamamoto, T., Davis, C.G., Brown, M.S., Schneider, W.J., Casey, M.L. *et al.* (1984) *Cell*, **39**, 27–38
43. McClelland, A., Kühn, L.C. and Ruddle, F.H. (1984) *Cell*, **39**, 267–74
44. Holland, E.C., Leungr, J.O. and Drickamer, K. (1984) *Proc. Natl. Acad. Sci.*, **81**, 7338–42
45. Kirchhausen, T., Harrison, S.C., Chow, E.P., Mattaliano, R.J. *et al.* (1987) *Proc. Natl. Acad. Sci.*, **84**, 8805–9
46. Kawakami, K., Noguchi, S., Noda, M., Takahashi, H., Ohta, T. *et al.* (1985) *Nature*, **316**, 733–6
47. Noguchi, S., Noda, M., Takahashi, H., Kawakami, K., Ohta, T. *et al.* (1986) *FEBS Lett.*, **196**, 315–20
48. Noda, M., Shimizu, S., Tanabe, T., Takai, T., Kayano, T. *et al.* (1984) *Nature*, **312**, 121–7
49. Tanabe, T., Takeshima, H., Mikami, A., Flockerzi, V., Takahashi, H. *et al.* (1987) *Nature*, **328**, 313–8
50. Schwarz, T.L., Tempel, B.L., Papazian, D.M., Jan., Y.N. and Jan, L.Y. (1988) *Nature*, **331**, 137–45
51. Loewenstein, W.R. (1987) *Cell*, **48**, 725–6
52. Warner, A. (1988) *J. Cell Sci.*, **89**, 1–7
53. Paul, D.L. (1986) *J. Cell Biol.*, **103**, 123–34
54. Hay, E.D. (1981) *J. Cell Biol.*, **91**, 205s–223s
55. Bornstein, P. and Sage, H. (1980) *Ann. Rev. Biochem.*, **49**, 957–1003
56. Lindhal, U. and Höök, M. (1978) *Ann. Rev. Biochem.*, **47**, 385–417
57. Hassell, J.R., Kimura, J.H. and Hascall, V.C. (1986) *Ann. Rev. Biochem.*, **55**, 539–67
58. Bourdon, M.A., Oldberg, A., Pierschbacher, M. and Ruoslahti, E. (1985) *Proc. Natl. Acad. Sci.*, **82**, 1321–5
59. Krusius, T. and Ruoslahti, E. (1986) *Proc. Natl. Acad. Sci.*, **83**, 7683–7
60. Sai, S., Tanaka, T., Kosher, R.A. and Tanzer, M.L. (1986) *Proc. Natl. Acad. Sci.*, **83**, 5081–5
61. Doege, K., Fernandez, P., Hassell, J.R., Sasaki, M. and Yamada, Y. (1986) *J. Bio. Chem.*, **261**, 8108–11
62. Krusius, T., Gehlsen, K.R. and Ruoslahti, E. (1987) *J. Biol. Chem.*, **262**, 13120–5
63. Yamada, K.M. (1983) *Ann. Rev. Biochem.*, **52**, 761–99
64. Pierschbacher, M.D. and Ruoslahti, E. (1984) *Nature*, **309**, 30–33
65. Yamada, K.M., Yamada, S.S. and Pastan, I. (1976) *Proc. Natl. Acad. Sci.*, **73**, 1217–21
66. Timpl, R., Rohde, H., Robey, P.G., Rennard, S.I., Foidart, J.M. and Martin, G.R. (1979) *J. Biol. Chem.*, **254**, 9933–7
67. Paulsson, M., Deutzmann, R., Timpl, R., Dalzoppo, D., Odermatt, E. and Engel, J. (1985) *EMBO J.*, **4**, 309–16
68. Wewer, U.M., Liotta, L.A., Jaye, M., Ricca, G.A., Drohan, W.N. *et al.*, (1986) *Proc. Natl. Acad. Sci.*, **83**, 7137–41
69. Terranova, V.P., Rohrbach, D.H. and Martin, G.R. (1980) *Cell*, **22**, 719–26
70. Smalheiser, N.R., Crain, S.M. and Reid, L.M. (1984) *Dev. Brain Res.*, **12**, 136–40
71. Sasaki, M., Kato, S., Kohno, K., Martin, G.R. and Yamada, Y. (1987) *Proc. Natl. Acad. Sci.*, **84**, 935–9
72. Grumet, M., Hoffman, S., Crossin, K.L. and Edelman, G.M. (1985) *Proc. Natl. Acad. Sci.*, **82**, 8075–9

73. Chiquet-Ehrischmann, R., Mackie, E.J., Pearson, C.A. and Sakakura, T. (1986) *Cell*, **47**, 131–9
74. Chiquet-Ehrismann, R., Kalla, P., Pearson, C.A., Beck, K. and Chiquet, M. (1988) *Cell*, **53**, 383–90

3 Cell surface markers and the immunoglobulin superfamily

Cell surface molecules expressed in restricted cell populations serve as cell surface markers. They are helpful in the identification and separation of these cells. Cell surface markers have been most extensively studied in lymphocytes and their precursors, probably because an understanding of lymphocytes is of critical importance in immunology. This chapter primarily deals with lymphocytes and their precursor cells.

Lymphocytes are classified into B lymphocytes and T lymphocytes. Both recognize foreign substances called antigens, and exert immunological defence against bacteria, viruses, fungi and neoplasias. After antigen recognition, B cells develop into plasma cells and secrete antibodies. T lymphocytes are classified into functional subsets according to their behaviour after antigen recognition. Helper T cells secrete helper factors, while suppressor T cells secrete suppressor factor(s); both regulate antigen formation positively or negatively. On the other hand, cytotoxic (killer) T cells kill the target cells carrying the antigen. Cell surface markers are invaluable in the analysis of lymphocyte subsets and precursors. Since lymphocytes must recognize antigens, there are antigen-recognizing molecules on their surface. These are cell surface immunoglobulins on B cells and T cell receptors on T cells. These molecules, which emerge at distinct stages during the differentiation of lymphocytes, will be discussed first. The immunoglobulin is the prototype of a large superfamily, the immunoglobulin superfamily, whose members are important not only in immunological recognition and as cell surface markers, but also in intercellular recognition during differentiation and development.

Table 3.1 Immunoglobulin classes

	Heavy (H) chain	Light (L) chain	Structure*	Molecular weight (K)
IgM	μ	κ or λ	$(\mu_2 L_2)_5$	900
IgD	δ	κ or λ	$\delta_2 L_2$	185
IgG	γ	κ or λ	$\gamma_2 L_2$	150
IgA	α	κ or λ	$(\alpha_2 L_2)_{1-2}$	160 320
IgE	ε	κ or λ	$\varepsilon_2 L_2$	200

*L = κ or λ.

3.1 IMMUNOGLOBULINS

Immunoglobulins (Ig) are the chemical entities of antibodies. They detect antigens and exert such effector functions as the initiation of complement-dependent cytolysis in antigen-bearing cells. Immunoglobulins are expressed only in B cell lineage, and cell surface immunoglobulins serve as markers of B cells. Secreted immunoglobulins are produced by the plasma cells which develop from B cells. There are several classes of immunoglobulins, such as IgG and IgM (Table 3.1) [1]. They are of course related to each other and IgG is representative of immunoglobulins.

An IgG molecule is composed of two light chains (λ or κ, 25 K) and two heavy chains (γ, 50 K) and is Y shaped. Two antigen binding sites are present in the N-terminal sides of IgG, while the effector function is carried out by the other half of the molecule (Figure 3.1). Limited papain digestion cleaves IgG into two monovalent antigen binding fragments (Fab) and one

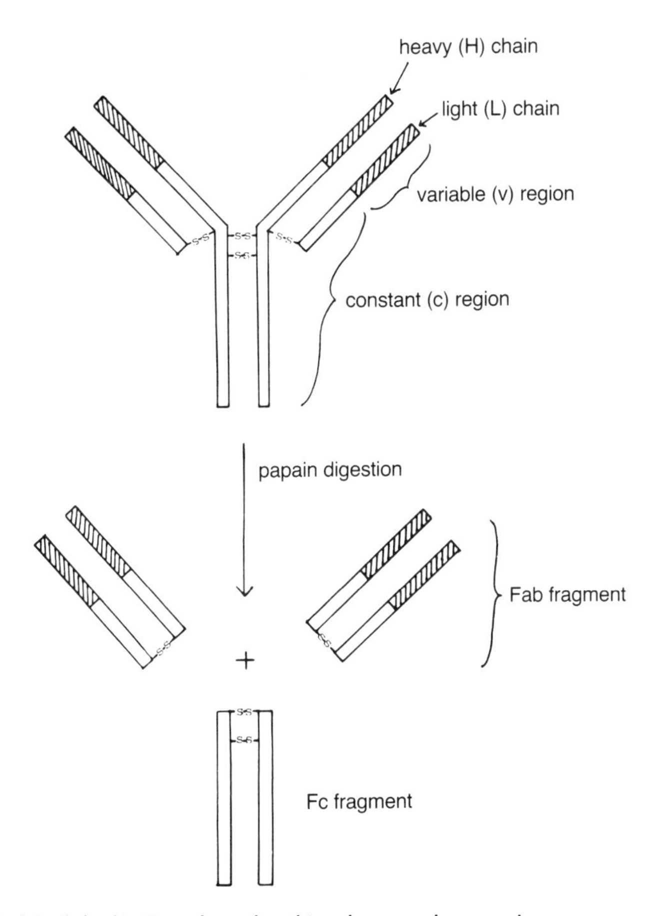

Figure 3.1 Model of IgG and mode of its cleavage by papain.

Fc fragment. Each polypeptide chain of immunoglobulin consists of a variable (V) region and constant (C) region [2, 3] (Figure 3.1). The V region variability forms the basis of antibody specificity. On the other hand immunoglobulin classes (IgG, IgM etc.) are determined by differences in the C region of heavy chains.

Immunoglobulins may be regarded as being formed by an assembly of domain structures. A light chain is composed of a V domain and a C domain. A heavy chain is composed of a V domain and 3 or 4 C domains [1, 2]. Extensive homologies are observed among sequences of V domains and among those of C domains. The sequences of a V domain and a C domain are distinct [2] but still have a weak degree of homology (Figure 3.2). In both the V domain and the C domain, polypeptides are basically folded into β-sheets, and form a typical three dimensional structure called the immunoglobulin fold. Thus, both immunoglobulin domains are believed to have evolved from a common ancestral molecule.

There are probably more than 10^6 molecules of immunoglobulins with different specificities; this corresponds to the enormous diversity of antigenic molecules which would be encountered. As the clonal selection theory predicted, the progeny of a B cell clone produce immunoglobulins of a certain specificity. A B lymphocyte expresses a cell surface immunoglobulin, whose specificity is identical to that of the immunoglobulin secreted by the plasma cells developed from the B lymphocyte clone. Antigen–antibody reactions occurring at the surface of B cells trigger their development to plasma cells.

The biochemical nature of cell surface immunoglobulins was revealed by Vitetta *et al.* [4]. They radioactively labelled splenic lymphocytes by the lactoperoxidase-I^{125} method. The lymphocytes were lysed in non-ionic detergent NP-40, and cell surface immunoglobulins in the lysate were precipitated employing anti-immunoglobulin sera. By analysing the immunoprecipitate using SDS polyacrylamide gel electrophoresis, the cell surface immunoglobulin has been shown to be a monomeric IgM. In addition to IgM, mature B cells also express cell surface IgD.

Subsequent studies have shown that the heavy chain of the cell surface IgM is slightly larger than that of secreted IgM, which is a pentamer of

V_H	18	V	K	V	S	C	K	A	S	G	G	T	F	S	R	S	A	I	I	W	- - - - - - - -	A	F	Y	F	C	96
$C_\gamma 1$	140	A	A	L	G	C	L	V	K	D	Y	F	P	E	P	V	T	V	S	W	- - - - - - - -	Q	T	Y	I	C	200

Figure 3.2 Comparison of V_H (CV domain of the heavy chain) and $C_\gamma 1$ (the first C domain of the heavy chain) of a myeloma IgG. Numbers indicate amino acid position. □; conserved amino acids; ⬚; conserved hydrophobic amino acids. (Based on [2].)

monomeric IgM. The cDNA cloning and sequencing of mRNA coding for IgM in lymphoma cells clarified the situation. The C-terminal sequence of the heavy chain of cell surface IgM has a transmembrane domain, while the soluble form does not have the domain. Two mRNA species specifying the heavy chain of the surface IgM and that of the soluble IgM are derived from the same precursor mRNA by differential splicing [5].

A DNA segment coding a V domain and a segment coding C domains are separated by a long intervening DNA segment in a germ line DNA sequence. During differentiation to B cells, immunoglobulin genes rearrange, and the V domain and the C domain sequences come closer. This rearrangement is required for the expression of Ig genes and generation of the diversity. The immunoglobulin initially synthesized by a B lymphocyte clone is IgM; rearrangement of the DNA specifying the C domain sequences result in switching to the production of other classes of immunoglobulins such as IgG and IgA. The expression of immunoglobulins and the status of the genes in B cell lineage is summarized in Table 3.2 [6].

3.2 T CELL RECEPTORS

T lymphocytes also specifically recognize antigens. From an analogy to cell surface immunoglobulins, the antigen recognizing molecule on the T cell surface was initially thought to be immunoglobulin. However, the presence of immunoglobulin on T cells remained controversial. An initial clue to the presence of a different class of molecules, T cell receptors (antigen receptor

Table 3.2 Rearrangements and expression of immunoglobulin genes during B lymphocyte development

Cells	Immunoglobulin genes rearranged	Immunoglobulins secreted or expressed
Stem cells	–	–
↓		
Pre B cells	μ	Intracellular μ chain
↓		
Resting B cells	μ, L	Cell surface IgM
↓		
B cells	μ, δ, L	Cell surface IgM plus cell surface IgD
↓		
IgM secreting plasma cells	μ, L	Secreted IgM
↓		
IgG (or IgA, IgE) secreting plasma cells	γ (or α, ε), L	Secreted IgG (or IgA, IgE)

of T cells), came from studies on 'clonotypic' antigens. A monoclonal antibody raised against a human cytotoxic T cell clone reacted only with this clone, and was called 'clonotypic'. The antibody specifically inhibited the cytotoxic activity of the T cell clone. The antigen recognized by the antibody was a protein (90 K) composed of two subunits, α (49 K) and β (43 K) [7]. Another clonotypic antibody reacting with another T cell clone also detected an antigen of 90 K. From these results, the 90 K molecules were concluded to be T cell receptors.

The protein structures of T cell receptors were revealed by recombinant DNA experiments. Yanagi *et al.* [8] and Hedrick *et al.* [9] isolated cDNA clones which were expressed in T cells but not in B cells. Furthermore, genomic DNA corresponding to the cDNA was found to be rearranged in T cell clones but not in B cells. From cDNA sequencing, the T cell specific clones have been shown to code transmembrane proteins which have both V domain-like and C domain-like sequences (Figure 3.3). Correlating the molecular biological data with the immunochemical data, it has been concluded that these cDNA clones code the β-chain of T cell receptors. Subsequently, the structure of the α-chain has been demonstrated; it also has V domain-like and C domain-like sequences. Thus, a T cell receptor (antigen receptor of T cells) has a domain structure somewhat similar to IgM (antigen receptor of B cell) (Figure 3.3). cDNA clones of T cell receptor γ/δ chains have been also isolated, this class of T cell receptor appears to function in a different way to α/β T cell receptors (section 3.3.1). In the fetal mouse, γ chain genes rearrange first, and the corresponding mRNA is also

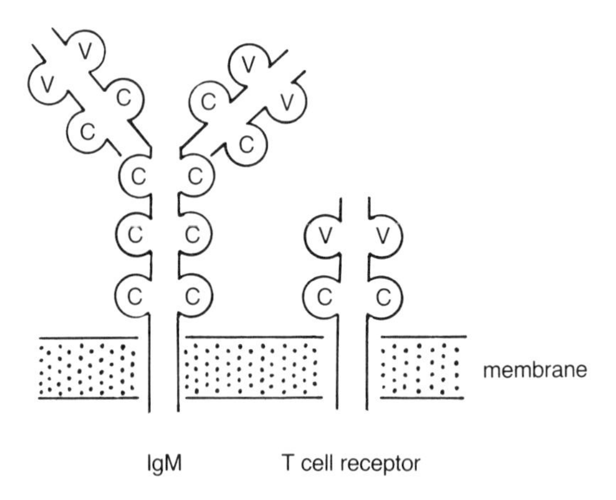

Figure 3.3 Comparison of domain structures of a membrane-bound IgM and a T cell receptor.

Table 3.3 Rearrangements and expression of T cell receptor genes during T cell development

Cells	Stem cell	Pro T cell	Pre T cell	T cell
(Gestation day)	(13 or earlier)	(14/15)	(16)	(17)
Rearrangement in the gene and expression of				
γ	–	+	+	+
β	–	–	+	+
α	–	–	–	+

Based on [10].

expressed earlier than α and β chains. The rearrangement and expression of the β-chain follows and then the α-chain gene is rearranged and expressed. On day 17 of gestation, both the α- and β-chain genes of T cell receptors are expressed, and this is the time when mature T cells become significantly detected. The status of the rearrangement and expression of T cell receptor genes is used to define the differentiation steps of T cells (Table 3.3) [10].

3.3 MAJOR HISTOCOMPATIBILITY COMPLEX

3.3.1 Class I molecules

The major histocompatibility complex (MHC) was discovered in the field of transplantation immunology. When mouse skin is grafted to a genetically different mouse, the skin is eventually rejected. More rapid rejection of the second graft indicates the important role of immunological mechanisms in the rejection phenomenon. Antigens involved in graft rejection are called histocompatibility antigens or transplantation antigens. Relying on the fact that a difference in a histocompatibility antigen results in the rejection of transplanted tumours, Snell produced many pairs of congenic mice strains by a series of backcrosses and selection using tumour grafts [11]. Generally speaking, a pair of congenic strains means that only a very limited number of genes are different in the pair. In the above cases only a gene specifying a histocompatibility antigen and some other genes near to it are expected to be different. Systematic studies utilizing congenic mice strains have concluded that H-2 antigen is the most potent histocompatibility antigen in the mouse [11–13]. Antisera raised by immunization between congenic mice strains which have different H-2 antigenic types react mostly with H-2 antigen; these sera have been valuable in the biochemical characterization of H-2 antigen. The human counterpart of H-2 antigen is HLA (human leukocyte antigen)-A, B, C. H-2 and HLA antigens are extremely polymorphic

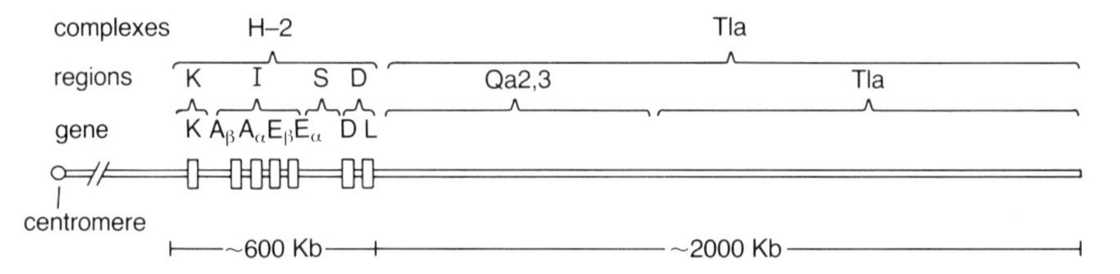

Figure 3.4 The major histocompatibility complex of the mouse. Organization of the gene complex in chromosome 17.

between different individuals. So far 100 or more different H-2 types (H-2 haplotypes; H-2^b, H-2^d, H-2^k etc.) are known. An individual mouse has three serologically distinct H-2 molecules, H-2D, H-2K and H-2L, and the molecular properties of the three H-2 antigens are similar. Genes for H-2 antigens are localized in a segment of chromosome 17 called the major histocompatibility complex (MHC) [13–15] (Figure 3.4). The biochemical nature of H-2 antigen was clarified in the 1970s by Nathenson and co-workers [16, 17], while that of HLA was elucidated by Strominger and co-workers [18, 19].

An H-2 molecule is a typical transmembrane protein, and can be solubilized from the membrane by limited papain digestion or by treatment with non-ionic detergents. The detergent solubilized antigen was immunoprecipitated and analysed by SDS polyacrylamide gel electrophoresis and the predominant polypeptide was revealed to be 45 K [20]. This turned out to be the heavy chain (α chain) of the antigen. An unexpected finding was that a light chain (β chain) of molecular weight 12 K was associated with the heavy chain of H-2 and HLA [21, 22]. The light chain was found to be identical to β_2-microglobulin, which had been isolated from the urine of patients with renal disorders and had been shown to be similar to the C domain of immunoglobulins (Figure 3.5). The finding of this immunoglobulin-like structure in H-2 and HLA was the beginning of the exciting story of the immunoglobulin superfamily.

The antigenic epitopes of H-2 and HLA antigens are in the heavy chain. The primary structures of the heavy chains were established initially by chemical methods for certain haplotypes, and for many other ones using recombinant DNA techniques. The extracellular domain of the heavy chain has been found to be composed of 3 domains, α_1, α_2 and α_3 [17]. α_1 and α_2 are the polymorphic domains and bear the antigenic sites. The α_3 domain has homology with the C domain of immunoglobulins (Figure 3.5), and is involved in association with the β-chain (Figure 3.6).

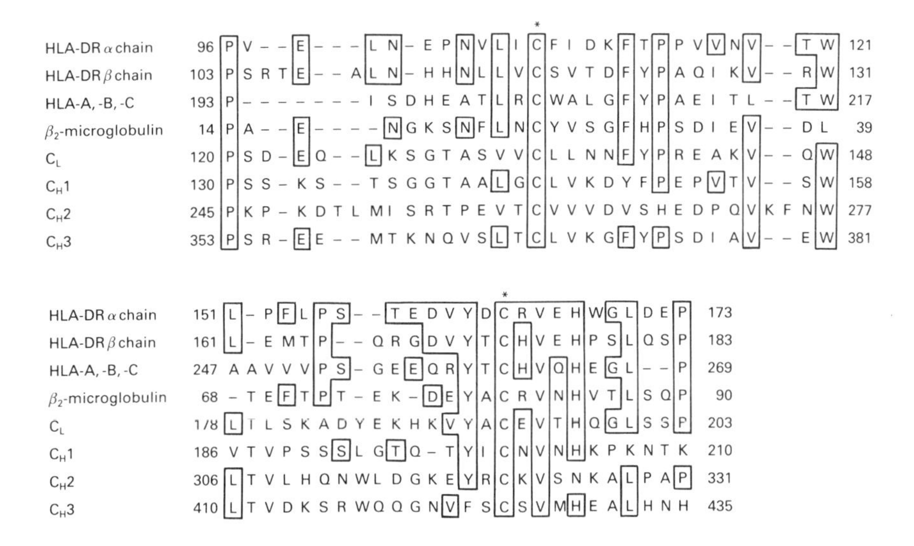

Figure 3.5 Comparison of sequences around two conserved cysteine residues (*) in HLA-DR, HLA-A, -B, -C, β_2-microglobulin and C domains of immunoglobulins (CL, CH1, CH2 and CH3). (Based on [34].)

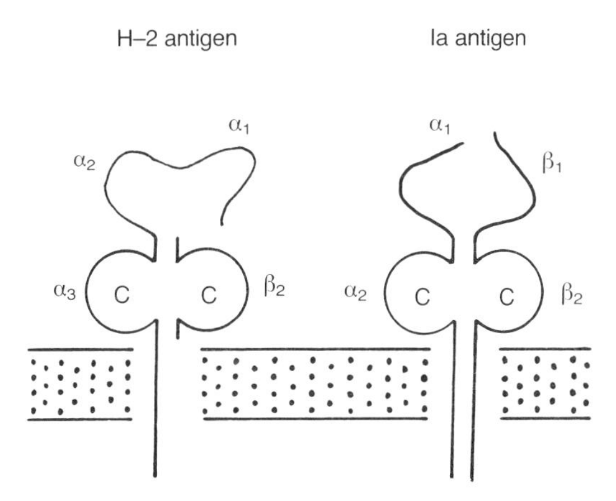

Figure 3.6 Domain structures of H-2 and Ia antigens. C: a domain similar to C domain of immunoglobulin.

X-ray crystallographic studies on HLA have revealed a groove in the α_1 and α_2 domain. The groove appears to recognize a foreign substance like an antibody [23]. Thus, H-2 (HLA) molecules and immunoglobulins are not only structurally related, but also functionally related. However, there is an important difference; extreme polymorphism of immunoglobulins is found within an individual, while the polymorphism of H-2 (HLA) is found between individuals. In other words, many different antigens are expected to be recognized by an H-2 (HLA) molecule.

A physiological role of H-2 (HLA) molecules has been clarified in relation to antigen recognition by cytotoxic T cells. When the T cell recognizes a foreign antigen on the target cell, the T cell and the target cell must have H-2 (HLA) antigen of an identical haplotype. For example, T cells from a $C_{57}BL/6$ mouse (H-2^b) infected with vaccinia virus, kill the virus infected cells of H-2^b haplotype, but not the cells of H-2^d haplotype (Figure 3.7). The phenomenon described by Zinkernagel is called H-2 (HLA) restriction [24].

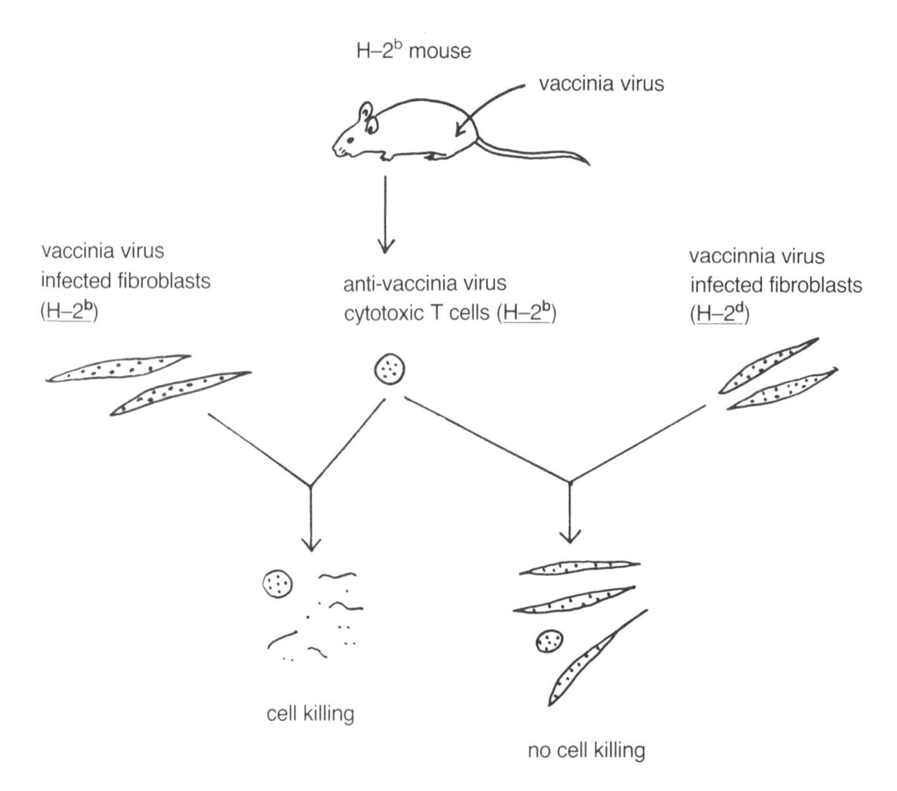

Figure 3.7 H-2 restriction of the action of cytotoxic T cells.

cifying Ia antigens with Ir gene promoted studies on Ia antigens [33]. lecular biological studies played crucial roles in clarifying the structure of antigens [34, 35]. Both the α chain and β chain are transmembrane lypeptides, and in each chain a domain located near to the nsmembrane domain is homologous to the C domain of immuno-bulins (Figure 3.5). The exterior portions of these molecules are poly-orphic and bear Ia antigenic sites (Figure 3.6).

An antigen usually stimulates two kinds of cells before the production an antibody. First are B lymphocytes, which are stimulated by antigen inding to the cell surface immunoglobulin. Secondly there are helper T ells, which are stimulated by antigen binding to the T cell receptor. The timulated T cells secrete helper factors (interleukins) which are required or the development of B cells to plasma cells (Chapter 4). The antigen to be ecognized by helper T cells is at first taken up by macrophages, processed nd the antigen fragment is presented to T cells. The Ia haplotypes must be dentical to enable the cooperation between macrophages and helper T cells [36]. This phenomenon was discovered in seminal form even before the finding of H-2 restriction [37, 38]. The role of Ia antigen in antigen presentation is now reasonably well understood; the role of Ia antigen on macrophages is quite similar to that of H-2 antigen on target cells (cf. section 3.7).

3.4 CD ANTIGENS

There are several differentiation antigens in CD (cluster of differentiation) nomenclature (see below). These antigens have been found to be antigens defining lymphocyte subsets. Boyse and co-workers relied on allogenic immunization (immunization between genetically different individuals in the same species) and found Ly2,3 (CD8), which is preferentially expressed in cytotoxic and suppressor T cells [39]. With the introduction of the monoclonal antibody technique, many monoclonal antibodies reacting with lymphocyte subsets could be prepared. The work of Reinherz *et al.* is

Table 3.4 Cells which are reactive with anti-CD antibodies (%)

Antibody	T cell subsets expressing the antigen in lymphocytes	Thymus cells	T cells (blood)
Anti-CD3	Total	10	100
Anti-CD4	Inducer	75	60
Anti-CD5	Total	10	100
Anti-CD8	Cytotoxic, suppressor	80	30

Based on [40].

Apparently, a T cell receptor on a cytotoxic T cell
only when it is complexed with self H-2 (HLA) antig
surface molecule of the T cell, CD8, adheres to H-2
target cell (cf. section 3.4).

H-2 antigen is expressed in most adult cells. How
significantly in EC cells 25 nor in embryonic ce
gestation [13]. The developmental control of H-2 a
significant interest, and protein factors binding to the
gene are under extensive study.

The segment of mouse chromosome 17, MHC, specil
molecules (TL and Qa), which have molecular weigl
antigen (45 K and 12 K). TL (thymus leukemia) antig
only on thymocytes and certain leukemic cells [26],
antigens are mostly restricted to cells of haematopoieti
structure of the heavy chains of TL and Qa antigens ha
those of H-2 antigens [28], and the light chain of TL and
form β_2-microglobulin. Therefore, H-2 (HLA), TL a
constitute a family; they are collectively called MHC Cl
simply Class I antigens. In the MHC of BALB/C mice, 36 C
been identified, of which 31 are in the Tla and Qa regions a
the H-2 region [14]. However, only a slight polymorp
individuals is found in TL antigen and in Qa antigens. Very
restricted antigen has been found to be recognized by the γ/δ
[29]. Thus Qa and TL antigens, which are collectively called M
molecules [30], appear to participate in different antigen
systems than H-2 antigens.

3.3.2 Class II molecules

A middle portion of MHC called the I region specifies cell surfa
different from Class I antigens (Figure 3.4). The antigens, called I
are heterodimers of 35–33 K (α chain) and 28–31 K (β chain) [
region is divided into I-A and I-E loci, and each one specifies a di
antigen [13, 15]. The human counterpart of Ia antigen is HLA-DR
DP [31]. Collectively, these antigens are called the Class II mol
MHC. They are expressed intensely on B cells, macrophages, dendı
and on thymic epithelium. Low levels of the antigens are also det
renal tubules and capillary wall endothelial cells. Ia antigen is impo
antigen presentation.

Ia antigen was initially found by specific sets of antisera rais
immunizing congenic mice strains. On the other hand, McDevitt *et al.*
the immune responsive (Ir) gene, which regulates the intensities of anti
production against synthetic polypeptides [32]. The coincidence of the

spe
Mc
Ia
po
tra
gl
m

o
b
c
s
f

especially impressive, since the antibodies they have developed not only distinguish lymphocyte subsets (Table 3.4), but also inhibit their specific functions [40].

For example, using T4 monoclonal antibody, they have separated the antigen-positive cells and negative cells. The antigen (CD4) positive cells are mostly helper cells and the antigen negative cells are mostly cytotoxic/suppressor cells [41]. Furthermore, T4 antibody inhibits the function of helper cells but not cytotoxic/suppressor cells in T cell mixtures.

The nomenclature of lymphocyte differentiation antigens was confusing for some time. CD (cluster of differentiation) nomenclature was introduced in 1984 to clarify the situation and it has become widely used. Among CD antigens, those listed in Table 3.5 are of special interest, and all will be described below except for CD1, which is similar to TL antigen already mentioned. With the aid of recombinant DNA techniques, primary structures have been established for CD antigens listed in the Table. Surprisingly, all the antigens in Table 3.5 are structurally related to immunoglobulins (Figure 3.8).

T lymphocytes form rosettes with sheep erythrocytes, and CD2 (50 K) [42] is the molecule involved in rosette formation. Anti-CD2 monoclonal antibody inhibits the cytotoxic action of T cells by inhibiting T cell adhesion to target cells. The ligand of CD2 on the target cell is LFA-3 (Chapter 5). Furthermore, on the T cell surface, a significant proportion of CD2 is present as a complex with CD3 [43], and probably participates in the signal transduction system of T cells (see below).

CD3 is a marker of mature T cells. The antigen is composed of 4 polypeptides (γ, 25 K; δ, 20 K; ε, 20 K; ζ, 16 K) and each polypeptide has a

Table 3.5 CD antigens useful for identification and separation of lymphocyte subsets

CD nomenclature	Original names		Typical leukocyte subpopulation expressing the antigen
	Man	Mouse	
CD1	T6/Leu6	TL	Corticothymocytes
CD2	T11/Leu5	–	All T cells forming E rosette
CD3	T3/Leu4	–	Mature T cells
CD4	T4/Leu3	L3T4	Subset of T cells, mostly helper
CD5	T1/Leu1	Ly-1 (Lyt-1)	All T cells plus subpopulation of B cells
CD8	T8/Leu2	Ly-2,3 (Lyt-2,3)	Subset of T cells, mostly cytotoxic/suppressor

Based on *Nature*, **325**, 660 (1987) and *Immunology Today*, **5**, 159 (1984).

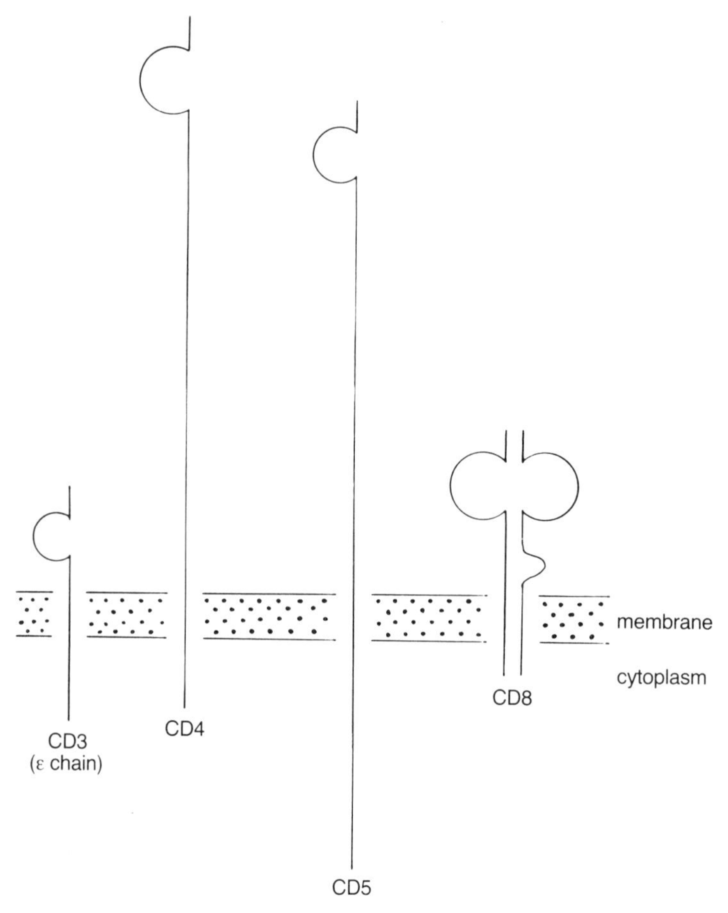

Figure 3.8 CD antigens have immunoglobulin-like domains (half circles).

domain of the immunoglobulin-like structure [44–46]. CD3 forms a complex with T cell receptors and is believed to be involved in signal transduction. When a T cell receptor recognizes an antigen, a conformational change induced in the receptor by the antigen is probably delivered to CD3 and activates phospholipase C, which is involved in the metabolism of phosphatidyl inositol [47]. Indeed, antibodies to CD3 mimic the action of specific antigens and cause a Ca^{2+} influx.

As mentioned before, CD4 (52 K) [48, 49] is expressed mostly in helper T cells, while CD8 (34 and 38 K) [50, 51] is expressed mostly in cytotoxic and suppressor T cells. These antigens are of special importance in the identification and separation of T cell subsets. Furthermore, monoclonal antibodies against CD4 and CD8 specifically inhibit the action of T cell subsets bearing the antigens. CD4 and CD8 are adhesive proteins; CD4 adheres to Class II antigen and CD8 to Class I antigen. CD4 might be also involved in signal transduction. A monoclonal antibody against CD4 inhibits the Ca^{2+} transport caused by antigen stimulation [52]. CD4 is also known as the binding site of HIV (human immunodeficiency virus) to target cells [53].

CD5 (67 K) is detected in all T cells and in a subset of B cells. The functional role of the B cell subset expressing CD5 is a matter of intense study. The antigen has an immunoglobulin-like domain [54, 55].

The differentiation of T cells from bone-marrow derived precursor cells occurs in the thymus. With the CD markers (cf. Tables 3.4 and 3.5), human thymocytes can be classified into 3 groups (Table 3.6). The cell in Stage I is the most undifferentiated one, while that in Stage III is the most differentiated one. As in the human thymocytes [40] CD4, CD8-double positive cells among mouse thymocytes are also believed to be precursors to mature thymocytes [56]. Thus, repeated injections of anti-CD8 antibody to the mouse blocks the development of $CD4^+$ cells.

3.5 OTHER CELL SURFACE MARKERS

3.5.1 Thy-1

Reif and Allen originally found Thy-1 antigen (called θ) by allogenic immunization in the mouse [57]. Thy-1 antigen is expressed in T cells but not in B cells [58]. Thus, in the mouse anti-Thy-1 antibodies have been used as a reagent to selectively destroy T cells with the aid of complement. Thy-1 antigen is also present in lymphocyte precursor cells, fibroblasts and nerve cells. In the human, its mode of expression is somewhat different, for

Table 3.6 A proposed pathway of intrathymic differentiation of human T cells

T cell subsets expressing the antigens in thymocytes	Thymus cells (%)	CD4/CD8	CD3/CD5
Stage I (early thymocytes) ↓	~10	–	–
Stage II (common thymocytes) ↓	>70	both +	–
Stage III (mature thymocytes) ↓	~10	only one +	both +

Based on [40].

example, it is highly expressed in the kidney. The antigen is composed of a single polypeptide chain of 17 K, and is anchored to plasma membranes by a phosphatidyl inositol glycan. The antigen also shows homology with the V domain of immunoglobulins [59] (cf. section 3.8).

Thy-1 antigen is probably involved in the control of cellular activity by cell surface signalling. Thy-1 specific antibodies have been found to elevate intracellular Ca^{2+} level and initiate the proliferation of murine T cells. The role of Thy-1 antigen has been also studied in transgenic mice, which are forced to express Thy-1 antigen in different tissues by an alteration of the promoter. Such mice developed a lymphoproliferative abnormality [60] or proliferative kidney disorder [61].

3.5.2 T200/B220 glycoprotein

Trowbridge *et al.* [62] found a difference in the cell surface proteins of T and B lymphocytes by surface labelling and SDS polyacrylamide gel electrophoresis. B lymphocytes have a characteristic molecule of 220 K (B220), while T lymphocytes have different molecules of 195 and 185 K (T200). Structural homology has been found between T200 and B220 and mRNA coding these glycoproteins may be produced by differential splicing [63]. Still another related molecule of 195–205 K (T200A) is detectable in proliferating T cells [64]. These molecules are expressed in the lymphoid cell lineages as shown in Figure 3.9. Furthermore, using a monoclonal antibody specifically reacting with B220, this glycoprotein has been shown to be expressed also in Pre B cells [65]. Thus, B220 is an excellent marker of B cell lineage. T-200 is also expressed in T cell precursors [64].

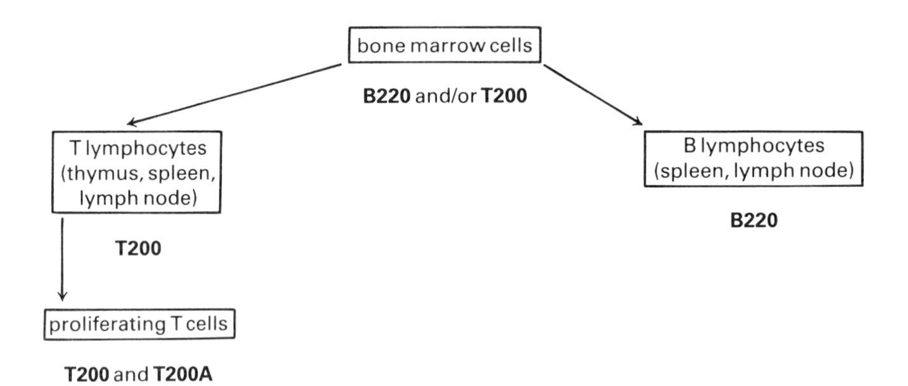

Figure 3.9 Expression of T200/B220 antigen in T and B lymphocytes and precursors. (Based on [64].)

3.6 APPLICATION OF CELL SURFACE MARKERS

Cell surface markers are indispensable for many studies in the area of immunology. Of special importance are cell surface Ig, B220, CD4, CD8 and Thy-1 (Tables 3.2, 3.4 and 3.7). A comprehensive list of cell surface markers useful in analysis of mouse haematopoieis is available [66].

A variety of methods are used to identify and separate lymphocyte subpopulations using antibodies reacting with cell surface markers. When the antibody, or the second antibody reacting with the first antibody, is labelled using a fluorescent dye, the antigen-positive cell can be seen under a fluorescence microscope. Using a fluorescence activated cell sorter (FACS), the degree of antigen expression may be described quantitatively. Furthermore, FACS enables the separation of the antigen positive cells from antigen negative cells. Other cell separation methods based on cell surface markers include killing of antigen-positive cells by the antibody and complement, and the adhesion of positive cells to antibody-coated (or fixed) particles.

Among the large number of applications for cell surface markers, the purification of mouse haematopoietic stem cells is an important example [67]. To isolate the stem cells, bone marrow cells were at first treated with fluoresceinated antibodies against CD4 and CD8. The antigen-positive cells were removed with the aid of magnetic beads coated with antibodies against fluorescein. Thy-1 positive cells were then isolated by a similar procedure. The Thy-1 positive cells were treated with antibodies against B220 (B cell lineage marker), Mac-1 (macrophage marker), Gr-1 (granulocyte marker) and Sca-1 (an antigen expressed in stem cells). By changing the second antibody, cells then expressing B220, Mac-1 and Gr-1 respectively were

Table 3.7 Cells expressing T200/B220 markers and other surface markers in lymphoid cell populations (%)

	Cells					
	Bone marrow	Thymus	Spleen	Lymph node	LPS*	Con A†
T200A	0	1–2	1–2	5–10	5–10	100
B220	30	0	50	40	95	1–2
CD8	0	70–80	10–15	10–15	5–10	70–85
Thy-1	4–5	98–100	35–45	55–65	5–10	95–100
T200 and B220	85–95	98–100	98–100	98–100	100	100

* Lipopolysaccharide-stimulated spleen cells.
† Con A-stimulated spleen cells.
Based on [64].

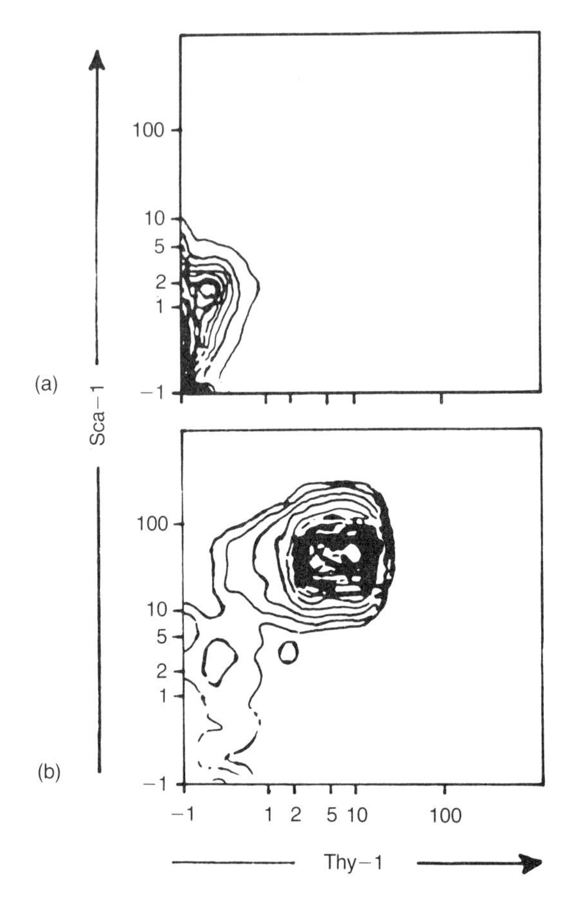

Figure 3.10 Identification of the isolated stem cell fraction by two-colour FACS. Whole bone marrow cells and isolated stem cells were comparatively analysed by two-colour FACS. (a) Whole bone marrow cells, (b) isolated stem cells. Direction of arrows indicates stronger antigenic expression. (Based on [67].) © 1988. AAAS.

stained by phycoerythrin, while Sca-1 positive cells were stained by Texas Red. Using FACS, a cell population with the following properties was isolated: intermediate level of Thy-1 expression, no expression of B220, Mac-1, Gr-1 and intense expression of Sca-1 (Figure 3.10). The isolated cells were believed to be enriched in haematopoietic stem cells, since only 30 of these cells were sufficient to save 50% of lethally-irradiated mice, and reconstitute all blood cell types in the survivors.

Cell surface markers intended for the useful analysis of other developmental systems are being developed in many laboratories. Perhaps other research areas where cell surface markers are especially helpful are

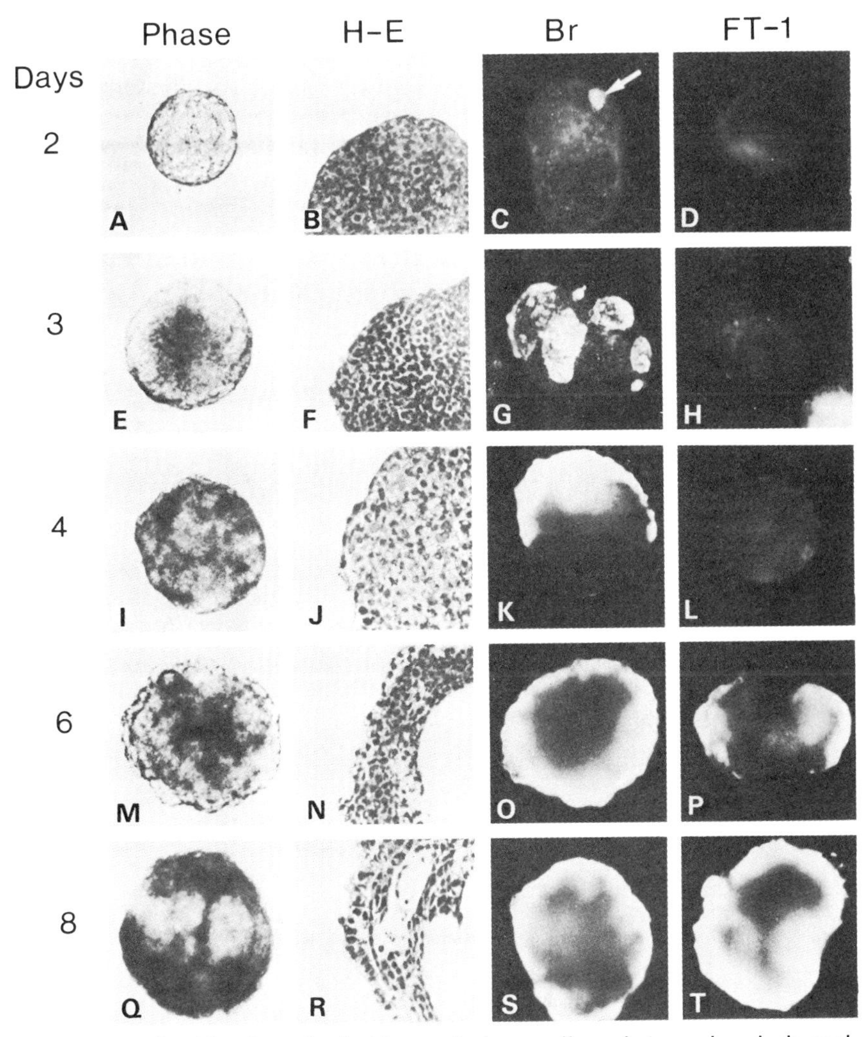

Figure 3.11 Identification of primitive endoderm cells and visceral endodermal cells differentiated from EC cells, by using two surface markers, brushin and FT-1. When aggregates of EC cells are treated with 10^{-7} M retinoic acid, the cells on the surface of the aggregate differentiate to visceral endoderm cells via primitive endoderm cells. The two markers distinguish the three cell populations as followsd: E cells: brushin (−), FT-1 (−); primitive endoderm cells: brushin (+), FT-1 (−); visceral endoderm cells: brushin (+), FT-1 (+). Immunofluorescence staining of cell aggregates are shown together with histochemical appearance of cell aggregates. Days: days treated with retinoic acid; Phase: phase contrast (A, E, I, M, Q); H-E: hematoxylin eosin staining (B, F, J, N, R); Br and FT-1: immunofluorescence microscopy stained by anti-brushin (C, G, K, O, S) and anti-FT-1 (D, H, L, P, T) antibodies, respectively. Arrow in C indicates the earliest sign of primitive endoderm differentiation. (Data taken from [68].) © Company of Biologists Ltd.

nerve differentiation and early mammalian embryogenesis. Some encouraging results have been reported. For example, two carbohydrate markers, galactocerebroside and A_2B_5 ganglioside are excellent markers of oligodendrocytes, astrocytes I/II and the precursors (Chapter 4). Two glycoprotein antigens, brushin and FT-1 are useful as markers to analyse visceral endoderm differentiation from EC cells (Figure 3.11) [68].

3.7 ROLE OF RECOGNITION BETWEEN T CELL RECEPTORS, MHC AND CD4/8 IN DIFFERENTIATION OF T CELLS

As described briefly in the previous section, not only T cell receptors, but also MHC Class I/Class II antigens and CD4/8 antigen are involved in antigen recognition by T cells. The currently accepted model of antigen recognition by cytotoxic/suppressor T cells is shown in Figure 3.12. At first, the antigen is complexed by the polymorphic portion of H-2/HLA antigen (MHC Class I antigen) on target cells. A T cell receptor recognizes the antigen plus the polymorphic portion of MHC Class I antigen, and the recognition signal is transduced into the cell with the aid of CD3. The recognition of the non-polymorphic portion of the MHC Class I molecule by CD8 intensifies the recognition reaction. Antigens presented by macrophages are recognized by helper T cells in a similar manner. In this case, the antigen fragment complexed with the polymorphic portion of

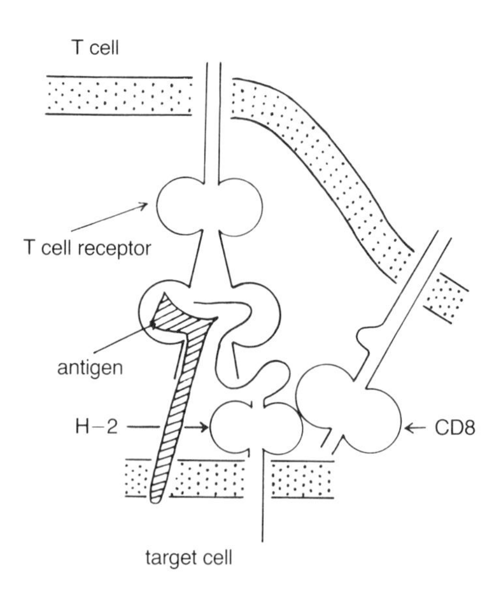

Figure 3.12 Model of antigen recognition by a cytotoxic T cell.

MHC Class II antigen on macrophages is believed to be recognized by T cell receptors, and CD4 on T cells reacts with the non-polymorphic portion of MHC Class II antigen.

This model of T cell recognition gives a clue in understanding the curious phenomenon of MHC restriction. That T cells and the target cells/macrophages must share the MHC haplotypes for T cell recognition implies that T cells have structures reacting with self MHC. The self-recognizing structures on T cells are T cell receptors themselves; they recognize the self MHC complexed with foreign antigens. Thus, among a number of T cell clones generated in the thymus, those recognizing self MHC (plus a foreign antigen) appear to survive. In addition, T cell clones recognizing self antigens (not MHC) will not survive. This process, called the education of T cells, is likely to occur during the generation of $CD4^+$, $CD8^-$ or $CD4^-$ and $CD8^+$ cells from CD4 and CD8 double positive cells. It has been proposed that this education is performed by recognition of self MHC on thymic epithelial cells by T cell receptors, which are present on T cell progenitor cells ($CD4^+$, $CD8^+$ cells). A precursor cell with a T cell receptor which recognizes self MHC Class I antigen, is believed to differentiate to $CD4^-$, $CD8^+$ cells, while a precursor cell with a T cell receptor which recognizes self MHC Class II antigen, is believed to differentiate to $CD4^+$, $CD8^-$ cell. Precursor cells with T cell receptors which do not react with self MHC antigens are likely to die out.

The following results obtained recently support the above mentioned mechanism. First, neonatal mice treated with anti-Class II monoclonal antibody fail to develop $CD4^+$, $CD8^-$ T cells [69]. Conversely, neonatal mice treated with anti-Class I monoclonal antibody fail to develop $CD4^-$, $CD8^+$ T cells [70]. Furthermore, a transgenic mouse expressing a T cell receptor, whose gene is isolated from a $CD4^-$, $CD8^+$ T cell clone, has a significantly higher amount of $CD4^-$, $CD8^+$ thymocytes (Figure 3.13) [71]. Since the T cell receptor from the $CD4^-$, $CD8^+$ T cell clone is expected to recognize self MHC Class I antigen, thymocytes with the receptor should select the $CD4^-$, $CD8^+$ pathway. This was indeed found to be the case.

The presence of both CD4 and CD8 antigens on the precursor cells may be the clue for the selection of $CD4^+$, $CD8^-$ or $CD4^-$, $CD8^+$ lineages. When the T cell receptor on the precursor cell recognizes self MHC Class I antigen, CD8 antigen probably recognizes the non-polymorphic portion of Class I antigen, and the recognition signal received by CD8 antigen appears to be important in adopting the $CD8^+$ lineage (Figure 3.14). The same reasoning is possible for the selection of the $CD4^+$ lineage.

As above, the recognition between T cell receptors, MHC and CD4/8 antigens are important not only in antigen recognition, but also, most probably, in determining the direction of differentiation of T cell subsets. However, the following point remains to be clarified. While it is proposed

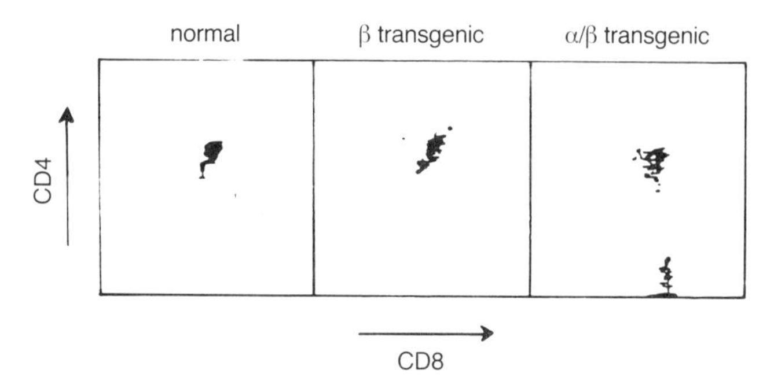

Figure 3.13 CD4 (−) CD8 (+) thymocyte population increased in a transgenic mouse expressing a T cell receptor (α/β) gene, which was isolated from CD4 (−) CD8 (+) T cells. The result of two-colour FACS analysis is presented. Direction of arrows indicates stronger antigenic expression. Normal: thymocytes from normal mice; β transgenic: thymocytesd from mice transgenic in T cell receptor β chain (no functional T cell receptor); α/β transgenic: thymocytes from mice transgenic in T cell receptor α/β gene (functional T cell receptor restricted with MHC Class I). (Based on [71].)

that a MHC molecule complexed with an antigen is recognized by a T cell receptor in antigen recognition, a free MHC molecule appears to be allowed to react with a T cell receptor in the education processes.

3.8 IMMUNOGLOBULIN SUPERFAMILY

The majority of lymphoid cell surface markers described in this chapter have domains homologous to the V or C domains of immunoglobulins. They are collectively called the immunoglobulin superfamily [72, 73]. Members of the superfamily also include Fc receptor [74, 75], poly-Ig receptor [76], MAG (myelin associated glycoprotein) [77], N-CAM (neural cell adhesion molecule) [78], L1 [79], CEA (carcinoembryonic antigen) [80], 1L-6 receptor [81], PDGF receptor and CSF-1 receptor [73]. Fc receptor is present on the surface of monocytes (macrophages), binds to the Fc region of IgG and triggers phagocytosis. Poly-Ig receptor is involved in the transport of IgA in epithelial cells. CEA is a tumour marker expressed in many carcinoma and embryonic tissues. N-CAM, L1 and MAG are cell adhesion proteins in the nervous system, while receptors for 1L-6, PDGF and CSF-1 are cell surface receptors involved in signal transduction.

Sequence homologies between members of the superfamily are most intense in sequences near two cysteine residues (Figures 3.5 and 3.15), which form an –S–S– bridge in immunoglobulins, and are thought to do so in other molecules. Notably, the sequences L (I, V)–X–C and Y (F)–X–C are

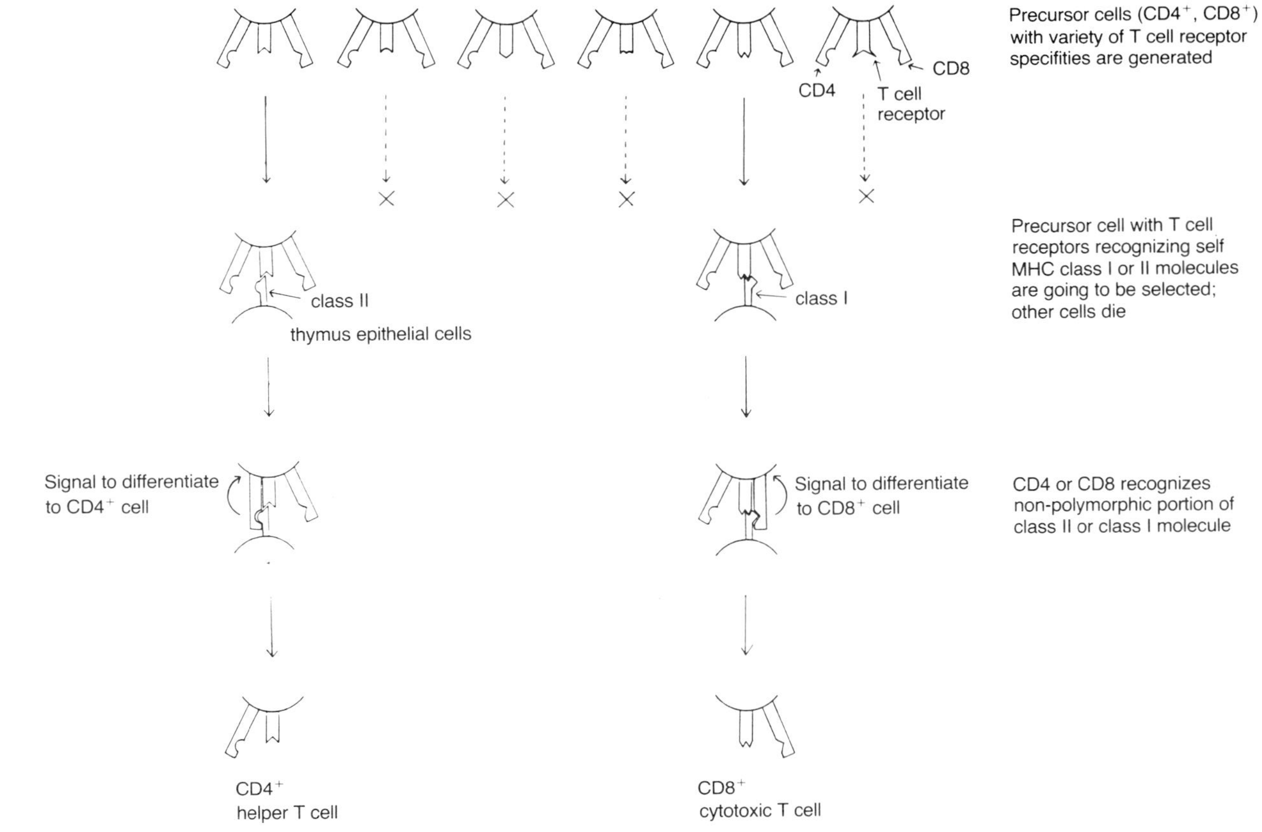

Figure 3.14 A model of ther role of MHC Class I/II and CD4/8 molecules in generation of CD4 (−) CD8 (+) or CD4 (+) CD8 (−) T cells.

found in all members of the superfamily. A tryptophan residue located about 14 residues downstream from the first cysteine residue is also well conserved.

The most prominent feature of the function of the immunoglobulin-like domains is that of domain–domain interaction, to form an inter- or intra-molecular complex. This interaction must be helpful in constructing immunoglobulins, T cell receptors, MHC Class I and II antigens and CD3, all of which are composed of subunits with immunoglobulin-like domains. Furthermore, the interaction is expected to be involved in intermolecular association by complex formation of T cell receptor with CD3; recognition of immunoglobulins by Fc receptor and poly Ig receptor and homophilic interaction of N-CAM.

The majority of the cell surface molecules involved in lymphoid cell recognition belong to the superfamily. Nerve cell recognition is probably another area where the immunoglobulin superfamily plays significant roles: N-CAM, myelin associated glycoprotein, L1, and Thy-1 have already been identified as involved in recognition in nerve cells.

Although in early embryonic cells most of the superfamily members mentioned above are not expressed significantly, we have recently identified two new members of the superfamily expressed in EC cells. The first one is embigin (formerly GP 70), which has a molecular weight of about 70 K and contains a V domain-like sequence (Figure 3.15) near the transmembrane domain [82]. Embigin is strongly expressed in mouse embryos before 10 days of gestation [83]. The second one is basigin (about 60 K), which has a domain homologous to the C domain of MHC Class II antigen [84]. Basigin is strongly expressed in several organs of the adult mice, including the testis. These results imply that interaction of early embryonic cells are partly mediated by mutual recognition of the superfamily members, many of which still remain to be detected. Such recognition in early embryos is likely to be important not only in cell adhesion (Chapter 5) but also in the regulation of cell differentiation; a prototype showing the latter role was described in section 3.7.

$V_{\kappa}1$	10	S L S A S V G D R V T I T C	23	73	L T I N S L Q P E D F A T Y Y C	88
Poly-Ig receptor	338	V L K G F P G G S V T I R C	351	405	V V L N Q L T A E D E G F Y W C	420
Thy-1	9	C L V N – – – Q N L R L D C	19	71	L T L A N F T T K D E G D Y F C	86
N-CAM	286	A V Y T W E G N Q V N I T C	299	337	L E V T P D S E N D F G N Y N C	352
MAG	334	T V V A V E G E T V S I L C	347	377	L E L P A V T P E D D G E Y W C	392
CEA	568	D S S Y L S G A N L N L S C	581	606	L F I A K I T P N N N G T Y A C	621
CD4	7	L V L G K E G E S A E L P C	20	75	L I I N K L K M E D S Q T Y I C	90
CD8	7	S L L V Q T N Q T A K M S C	20	79	L K I M D V K P E D S G F Y F C	94
Embigin	169	S L I A Y V G D S T V L K C	182	225	L K I K H L L E E D G G S Y W C	240

Figure 3.15 Comparison of conserved sequences in members of the immunoglobulin superfamily.

3.9 T/t GENETIC REGION

T/t genetic region of the mouse [85–87] is mentioned here, since it has often been discussed in relation to MHC. T locus was identified through a series of dominant mutations causing short tail [88]. Homozygous T mutant is embryonic lethal; embryogenesis is arrested after the primitive streak stage (Figure 3.16). t-Haplotypes were found to be genetic factors, which in heterozygous T/t mouse, cause complete absence of a tail. In the majority of cases, a homozygous t-haplotype is embryonic lethal; the stages where embryogenesis is arrested are different according to t-haplotypes (Figure 3.16). T locus and t-haplotypes are located in mouse chromosome 17. Although T locus is probably a defined genetic locus near the centromere, t-haplotypes cover a wide segment of chromosome 17 (12–15 cM; $20–30 \times 10^3$ kilobases) which encompasses T as well as the entire MHC region. In t-haplotypes, two inversions occur in chromosome 17, so that recombination with the wild type chromosome is severely inhibited in its long range [89, 90] (Figure 3.17). Because of the inhibition of recombination, several sets of genes are kept in the chromosome region.

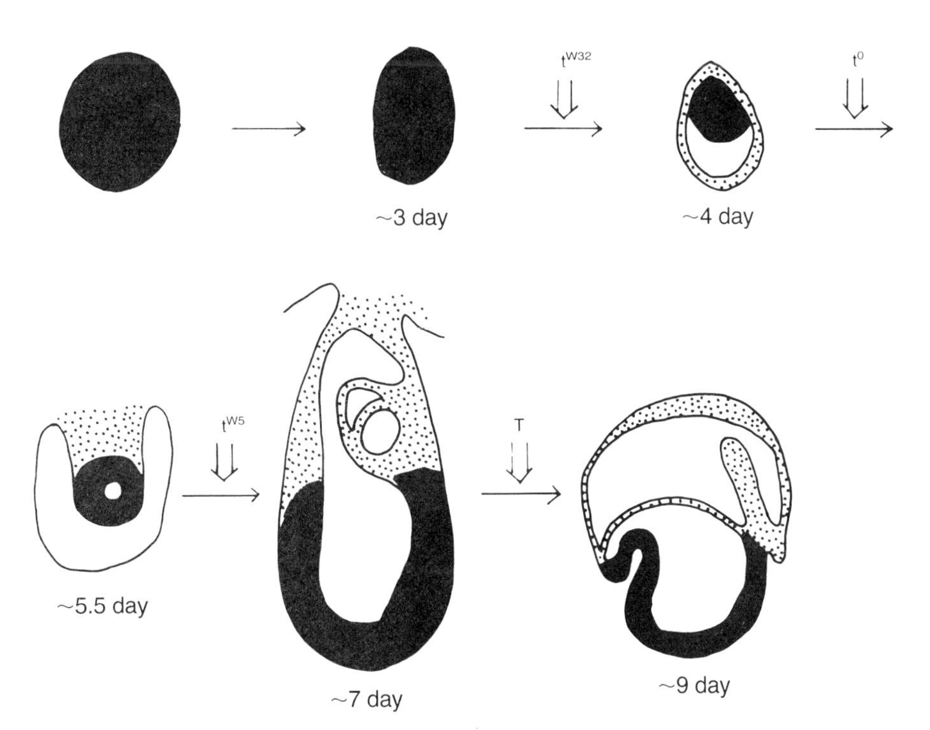

Figure 3.16 Arrest of embryogenesis in mice carrying homozygous T/t factors (based on [85]).

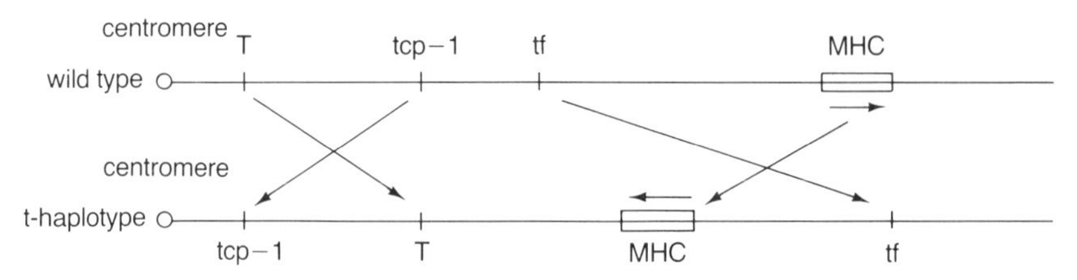

Figure 3.17 Two-fold inversion of chromosome 17 in mice with t-haplotypes.

Some are t lethal factors mentioned above. Also important are factor(s) responsible for the phenomenon called transmission ratio distortion. When a male mouse carrying a t-haplotype (t/+) is mated with a wild mouse, usually the offspring is t/+. On the other hand, when a female mouse with a t-haplotype (t/+) is mated with a wild mouse, t/+ and +/+ offsprings are produced at the ratio of 1 : 1. Transmission ratio distortion helps to maintain t-haplotypes in mice populations for a long period in spite of the recessive lethal nature of the haplotypes. Transmission ratio distortion is likely to be specified by genes involved in sperm–egg interaction, and the gene product is probably expressed on the surface of sperm. Genes probably involved in transmission ratio distortion have been cloned and characterized [91, 92, 93].

Several researchers have proposed that, as well as T locus, t lethality factors code for the cell surface proteins required for embryonic cell interactions. Presence of MHC in the region of t-haplotypes might support the proposal. Furthermore, T/T embryonic cells are more adhesive than wild type cells [94].

Genetic analysis has indicated that t^{w5} lethality factor is present within 250 Kb of H-2K. Bennett and co-workers cloned ~ 170 Kb of DNA containing H-2K and revealed the presence of 5 additional genes in the region [95]. Four of these are abundantly expressed in embryos, and the fifth is exclusively expressed in lymphoid cells. It is possible that one of the four embryo-expressed genes is the wild type gene of the t^{w5} factor. Furthermore, some of the embryo-expressed genes may specify new members of the immunoglobulin superfamily involved in embryonic cell interactions.

REFERENCES

1. Watson, J.D., Hopkins, N.H., Roberts, J.W., Steitz, J.A. and Weiner, A.M. (1987) *Molecular Biology of the Gene*, The Benjamin/Cummings Publishing Co., Menlo Park, USA
2. Edelman, G.M., Cunningham, B.A., Gall, W.E., Gottlieb, P.D., Rutishauser, U. and Waxdal, M.J. (1969) *Proc. Natl. Acad. Sci.*, **63**, 78–85
3. Wu, T.T. and Kabat, E.A. (1970) *J. Exp. Med.*, **132**, 211–50

4. Vitetta, E.S., Baur, S. and Uhr, J.W. (1971) *J. Exp. Med.*, **134**, 242–64
5. Rogers, J., Early, P., Carter, C., Calame, K., Bond, M., Hood, L. and Wall, R. (1980) *Cell*, **20**, 303–12
6. Rogers, J. and Wall, R. (1984) *Advan. Immunol.*, **35**, 39–59
7. Meuer, S.C., Fitzgerald, K.A., Hussey, R.E., Hodgdon, J.C., Schlossman, S.F. and Reinherz, E.L. (1983) *J. Exp. Med.*, **157**, 705–19
8. Yanagi, Y., Yoshikai, Y., Leggett, K., Clark, S.P., Aleksander, I. and Mak, T.W. (1984) *Nature*, **308**, 145–9
9. Hedrick, S.M., Cohen, D.I., Nielsen, E.A. and Davis, M.M. (1984) *Nature*, **308**, 149–53
10. Haars, R., Kronenberg, M., Gallatin, W.M., Weissman, I.L., Owen, F.L. and Hood, L. (1986) *J. Exp. Med.*, **164**, 1–24
11. Snell, G. (1981) *Science*, **213**, 172–8
12. Klein, J. (1979) *Science*, **203**, 516–21
13. Flavell, R.A., Allen, H., Burkly, L.C., Sherman, D.H., Waneck, G.L. and Widera, G. (1986) *Science*, **233**, 437–43
14. Winoto, A., Steinmetz, M. and Hood, L. (1983) *Proc. Natl. Acad. Sci.*, **80**, 3425–9
15. Steinmetz, M., Minard, K., Horvath, S., McNicholas, J., Srelinger, J. *et al.* (1982) *Nature*, **300**, 35–42
16. Nathenson, S.G., Uehara, H., Ewenstein, B.M., Kindt, T.J. and Coligan, J.E. (1981) *Ann. Rev. Biochem.*, **50**, 1025–52
17. Coligan, J.E., Kindt, T.J., Uehara, H., Martinko, J. and Nathenson, S.G. (1981) *Nature*, **291**, 35–9
18. Orr, H.T., López de Castro, J.A., Lancet, D. and Strominger, J.L. (1979) *Biochemistry*, **18**, 5711–20
19. López de Castro, J.A., Bragado, R., Strong, D.M. and Strominger, J.L. (1983) *Biochemistry*, **22**, 3961–9
20. Schwartz, B.D. and Nathenson, S.G. (1971) *J. Immunol.*, **107**, 1363–7
21. Cresswell, P., Turner, M.J. and Strominger, J.L. (1973) *Proc. Natl. Acad. Sci.*, **70**, 1603–7
22. Grey, H.M., Kubo, R.T., Colon, S.M., Poulik, M.D., Cresswell, P. *et al.* (1973) *J. Exp. Med.*, **138**, 1608–12
23. Bjorkman, P.J., Saper, M.A., Samraoui, B., Bennett, W.S., Strominger, J.L. and Wiley, D.C. (1987) *Nature*, 329, 506–12
24. Zinkernagel, R.M. and Doherty, P.C. (1979) *Advan. Immunol.*, **27**, 51–177
25. Morello, D., Daniel, F., Baldacci, P., Cayre, Y., Gachelin, G. and Kourilsky, P. (1982) *Nature*, **296**, 260–2
26. Boyse, E.A. and Old, L.J. (1971) *Transplantation*, **11**, 561–2
27. Stanton, T.H. Cathins, C.E., Jandinski, J., Schendel, O., Cantor, S.H. and Boyse, E.A. (1976) *J. Exp. Med.*, **148**, 963–73
28. McKenzie, I.F.C. and Potter, T. (1979) *Advan. Immunol.*, **27**, 179–337
29. Vidovič, D., Roglič, M., McKune, K., Guerder, S., Mackay, C. and Dembić, Z. (1989) *Nature*, **340**, 646–50
30. Strominger, J.L. (1989) *Cell*, **57**, 895–8
31. Möller, E., Carlsson, B. and Wallin, J. (1985) *Immunol. Rev.*, **85**, 107–28
32. McDevitt, H.O. and Sela, M. (1965) *J. Exp. Med.*, **122**, 517–31
33. McDevitt, H.O., Shreffler, D.C. and Stimpfling, J.H. (1969) *J. Clin. Invest.*, **48**, 57a
34. Larhammar, D., Gustafsson, K., Claesson, L., Bill, P. and Wiman, K. *et al.* (1982) *Cell*, **30**, 153–61

35. McNicholas, J., Steinmetz, M., Hunkapiller, T., Jones, P. and Hood, L. (1982) *Science*, **218**, 1229–32
36. Thomas, D.W., Yamashita, U. and Shevach, E.M. (1977) *J. Immunol.*, **119**, 223–6
37. Katz, D.H., Hamaoka, T., Dorf, M.E., Benacerraf, B. (1973) *Proc. Natl. Acad. Sci.*, **70**, 2624–8
38. Shevach, E.M. and Rosenthal, A.S. (1973) *J. Exp. Med.*, **138**, 1213–29
39. Boyse, E.A., Miyazawa, M., Aoki, T. and Old, L.J. (1968) *Proc. Roy. Soc. Ser. B.*, **170**, 175–93
40. Reinherz, E.L. and Schlossman, S.F. (1980) *Cell*, **19**, 821–7
41. Reinherz, E.L., Kung, P.C., Goldstein, G. and Schlossman, S.F. (1979) *Proc. Natl. Acad. Sci.*, **76**, 4061–5
42. Seed, B. and Aruffo, A. (1987) *Proc. Natl. Acad. Sci.*, **84**, 3365–9
43. Brown, M.H., Cantrell, D.A., Brattsand, G., Crumpton, M.J. and Gullberg, M. (1989) *Nature*, **339**, 551–3
44. Gold, D.P., Clevers, H., Alarcon, B., Dunlap, S., Novotny, J. *et al.* (1987) *Proc. Natl. Acad. Sci.*, **84**, 7649–53
45. van den Elsen, P., Shepley, B.-A., Borst, J., Coligan, J.E., Markham, A.F. *et al.* (1984) *Nature*, **312**, 413–8
46. Haser, W.G., Saito, H., Koyama, T. and Tonegawa, S. (1987) *J. Exp. Med.*, **166**, 1186–91
47. Imboden, J.B. and Stobo, J.D. (1985) *J. Exp. Med.*, **161**, 446–56
48. Maddon, P.J., Littman, D.R., Godfrey, M., Maddon, D.E., Chess, L. and Axel, R. (1985) *Cell*, **42**, 93–104
49. Tourvieille, B., Gorman, S.D., Field, E.H., Hunkapiller, T. and Parnes, J.R. (1986) *Science*, **234**, 610–14
50. Johnson, P. and Williams, A.F. (1986) *Nature*, **323**, 74–6
51. Nakauchi, H., Shinkai, Y. and Okumura, K. (1987) *Proc. Natl. Acad. Sci.*, **84**, 4210–14
52. Rosoff, P.M., Burakoff, S.J. and Greenstein, J.L. (1987) *Cell*, **49**, 845–53
53. Dalgleish, A.G., Beverly, P.C.L., Clapham, P.R., Crawford, D.H., Greaves, M.F. and Weiss, R.A. (1984) *Nature*, **312**, 763–7
54. Jones, N.H., Clabby, M.L., Dialynas, D.P., Huang, H.S., Herzenberg, L.A. and Strominger, J.L. (1986) *Nature*, **323**, 346–9
55. Huang, H.S., Jones, N.H., Strominger, J.L. and Herzenberg, L.A. (1987) *Proc. Natl. Acad. Sci.*, **84**, 204–8
56. Smith, L. (1987) *Nature*, **326**, 798–800
57. Reif, A.E. and Allen, J.M.V. (1964) *J. Exp. Med.*, **120**, 413–33
58. Raff, M.C. (1969) *Nature*, **224**, 378–9
59. Williams, A.F. and Gagnon, J. (1982) *Science*, **216**, 696–703
60. Chen, S., Botteri, F., van der Putten, H., Landel, C.P. and Evans, G.A. (1987) *Cell*, **51**, 7–19
61. Kollias, G., Evans, D.J., Ritter, M., Beech, J., Morris, R. and Grosveld, F. (1987) *Cell*, **51**, 21–31
62. Trowbridge, I.S., Ralph, P. and Bevan, M.J. (1975) *Proc. Natl. Acad. Sci.*, **72**, 157–61
63. Raschke, W.C. (1987) *Proc. Natl. Acad. Sci.*, **84**, 161–5
64. Sarmiento, M., Loken, M.R., Trowbridge, I.S., Coffman, R.L. and Fitch, F.W. (1982) *J. Immunol.*, **128**, 1676–84
65. Coffman, R.L. and Weissman, I.L. (1981) *Nature*, **289**, 681–3
66. Holmes, K.L. and Morse III, H.C. (1988) *Immunology Today*, **9**, 344–50

67. Spangrude, G.J., Heimfeld, S. and Weissman, I.L. (1988) *Science*, **241**, 58–62
68. Sato, M., Ozawa, M., Hamada, H., Kasai, M., Tokunaga, T. and Muramatsu, T. (1985) *J. Embryol. Exp. Morphol.*, **88**, 165–82
69. Kruisbeek, A.M., Fultz, M.J., Sharrow, S.O., Singer, A. and Mond, J.J. (1983) *J. Exp. Med.*, **157**, 1932–46
70. Marušić-Galěsić, S., Stephany, D.A., Longo, D.L. and Kruisbeek, A.M. (1988) *Nature*, **333**, 180–3
71. Teh, H.S., Kisielow, P., Scott, B., Kishi, H., Uematsu, Y. *et al.* (1988) *Nature*, **335**, 229–33
72. Hood, L., Kronenberg, M. and Hunkapiller, T. (1985) *Cell*, **40**, 225–9
73. Williams, A.F. and Barclay, A.N. (1988) *Ann. Rev. Immunol.*, **6**, 381–406
74. Lewis, V.A., Koch, T., Plutner, H. and Mellman, I. (1986) *Nature*, **324**, 372–5
75. Ravetch, J.V., Luster, A.D., Weinshank, R., Kochan, J., Pavlovec, A. *et al.* (1986) *Science*, **234**, 718–25
76. Mostov, K.E., Friedlander, M. and Blobel, G. (1984) *Nature*, **308**, 37–43
77. Arquint, M., Roder, J., Chia, L., Down, J., Wilkinson, D. *et al.* (1987) *Proc. Natl. Acad. Sci.*, **84**, 600–4
78. Cunningham, B.A., Hemperly, J.J., Murray, B.A., Prediger, E.A., Brackenbury, R. and Edelman, G.M. (1987) *Science*, **236**, 799–806
79. Moos, M., Tacke, R., Scherer, H., Teplow, D., Früh, K. and Schachner, M. (1988) *Nature*, **334**, 701–3
80. Oikawa, S., Nakazato, H. and Kosaki, G. (1987) *Biochem. Biophys. Res. Commun.*, **142**, 511–18
81. Yamasaki, K., Taga, T., Hírata, Y., Yawata, H., Kawanishi, Y. *et al.* (1988) *Science*, **241**, 825–8
82. Ozawa, M., Huang, R.-P., Furukawa, T. and Muramatsu, T. (1988) *J. Biol. Chem.*, **263**, 3059–62
83. Huang, R.-P. (1990) PhD Thesis, Kagoshima University, in press.
84. Miyauchi, T., Kanekura, T., Yamaoka, A., Ozawa, M., Miyazawa, S. and Muramatsu, T. (1989) *J. Biochem.*, **107**, 316–23
85. Silver, L.M. (1985) *Ann. Rev. Genet.*, **19**, 179–208
86. Gachelin, G. (1988) *La Recherche*, **19**, 152–61
87. Klein, J. (1975) *Biology of the mouse histocompatibility-2-complex*, Springer Verlag, Berlin
88. Dobrovoloskaia-Zavadskaia, N. and Kobozieff, N. (1932) *C. R. Soc. Biol.*, **110**, 782–4
89. Artzt, K., Shin, H. and Bennett, D. (1982) *Cell*, **28**, 471–6
90. Herrmann, B., Bućan, M., Mains, P.E., Frischauf, A.-M., Silver, L.M. and Lehrach, H. (1986) *Cell*, **44**, 469–76
91. Willison, K.R., Dudley, K. and Potter, J. (1986) *Cell*, **44**, 727–38
92. Rappold, G.A., Stubbs, L., Labeit, S., Ozkvenjakov, R.B. and Lehrach, H. (1987) *EMBO J.*, **6**, 1975–80
93. Schimenti, J., Cebra-Thomas, J.A., Decker, C.L., Islam, S.D., Pilder, S.H. and Silver, L.M. (1988) *Cell*, **55**, 71–8
94. Yanagisawa, K.O. and Fujimoto, H. (1977) *J. Embryol. Exp. Morphol.*, **40**, 277–83
95. Abe, K., Wei, J.-F., Wei, F.-S., Hsu, Y.-C., Uehara, H. *et al.* (1988) *EMBO J.*, **7**, 3441–9

4 Growth factors and receptors

Growth factors are polypeptides that promote growth of certain cell populations (Table 4.1). Nerve growth factor (NGF), which promotes the outgrowth of neurites from ganglia, is the first growth factor isolated [1]. Cohen and Levi-Montalcini achieved the isolation by using mouse submandibular salivary gland as the source. During the course of studies on NGF, Cohen found a different factor promoting the growth and keratinization of epidermal tissue. This factor, called epidermal growth factor (EGF), has been revealed to be a polypeptide (6 K) with characteristic intrachain –S–S– bridges (Figure 4.1) [2]. Since then growth factors acting on a spectrum of cells have been isolated and characterized. Of the various growth factors described so far (cf. Table 4.1), the following groups of factors need special attention because of their profound effects on cell differentiation:

Haematopoietins stimulate the differentiation of blood cells by promoting the growth of the intermediate cells of differentiation [3–5]. Erythropoietin is the prototype of haematopoietins. Erythropoietin is secreted by the kidney when oxygen levels are low, and promotes the differentiation of erythrocytes. Miyake and Goldwasser isolated erythropoietin from 2550 l of human urine; the purified factor is a protein of 39 K [6].

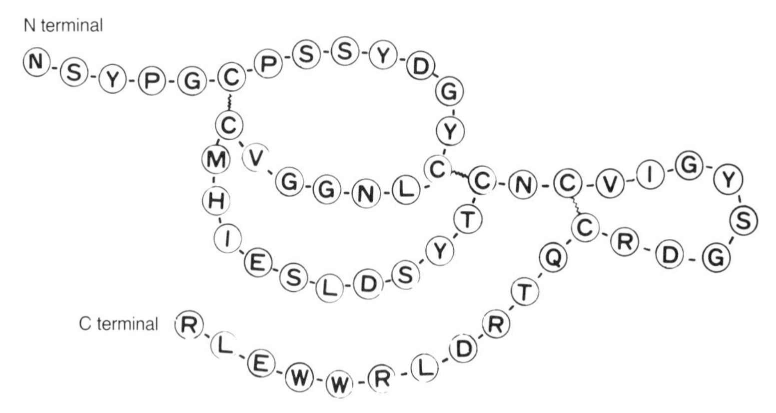

Figure 4.1 Structure of epidermal growth factor (EGF).

Table 4.1 Representative growth factors

Name (abbreviations)	Molecular weight (K)	Example of origin	Example of target cells	Reference
Nerve growth factor (NGF)	$(13)_2$	Salivary gland	Nerve cells	57
Epidermal growth factor (EGF)	6	Salivary gland	Epidermal cells	2
Fibroblast growth factor (FGF)	16 and 14.5 (basic one)	Brain	Fibroblasts	58
Platelet-derived growth factor (PDGF)	$(\sim 15)_2$	Platelet	Fibroblasts	59
Transforming growth factor-β (TGF-β)	$(12.5)_2$	Tumour cells Placenta	Mesenchyme cells	12
Granulocyte–macrophage colony stimulating factor (GM-CSF)	23	Lung	Granulocyte–macrophage progenitor cells	60
Interleukin 1 (IL-1)	18	Macrophages Keratynocytes	T cells Fibroblasts	61
Interleukin 2 (IL-2)	15	Helper T cells	T cells	62
Interleukin 3 (IL-3)	28	T cells	Haematopoietic cells	63, 64
Interleukin 4 (IL-4)	20	T cells	B cells Haematopoietic cells	19, 65
Interleukin 5 (IL-5)	$(23)_2$	T cells	B cells	8, 9
Interleukin 6 (IL-6)	21	T cells Fibroblasts	B cells	66, 67 68

Metcalf cultured bone marrow cells in soft agar and found that stem cells present in the bone marrow cells differentiated into granulocytes and macrophages. For such differentiation, the cells need a haematopoietin called GM-CSF (granulocyte-macrophage colony stimulating factor), which has been isolated from the lung of mice injected with endotoxin. The purified factor is a protein of molecular weight 23 K [7]. In addition, a factor promoting only macrophage differentiation (M-CSF or CSF-1) and a factor promoting only granulocyte differentiation (G-CSF) are known.

Leukocytes, especially T cells and monocytes, secrete factors which promote the growth and function of other leukocytes. These factors are collectively called interleukins (IL). Interleukins frequently have the function of haematopoietins, and distinctions between interleukins and haematopoietins are no longer present. For example, a factor promoting eosinophil differentiation [8] is identical to a B cell growth factor called IL-5 [9].

Normal fibroblasts cannot grow in soft agar, while transformed fibroblasts can. This property correlates well with the tumour-forming capability of transformed fibroblasts in nude mice. Todaro found that the culture medium of transformed fibroblasts contained factors that promoted the growth of normal fibroblasts in soft agar [10]. These factors are collectively called transforming growth factors (TGF). TGF-α resembles EGF in structure. TGF-β is unrelated to EGF, but it also has a characteristic shape due to intramolecular –S–S– bridges [11, 12]. TGFs are secreted not only by transformed fibroblasts but also by other cells, especially certain embryonic cells.

The important role of growth factors in the regulation of cellular activities has been indicated by the fact that the products of many oncogenes have homology with growth factors, their receptors and other components of the signal transduction system. The seminal finding is the homology of *sis* oncogene of simian sarcoma virus with platelet-derived growth factor (PDGF) [13]. Subsequently, the product of *int-2* proto-oncogene, which is activated by the insertion of mammary tumour virus [14], and the product of *hst*, which is an oncogene frequently activated in human stomach carcinoma, have been found to be homologous with fibroblast growth factor [15]. Furthermore, tyrosine kinase activity which is present in cytoplasmic domains of several growth factor receptors was initially found in the products of viral oncogenes such as *src*. EGF receptor, which has a tyrosine kinase domain, is especially homologous to *erb* B oncogene of avian erythroblastosis virus (Chapter 2).

Growth factors were implicated in cell differentiation first in mammalian systems, notably blood cell differentiation. Then, developmental genetics in *Drosophila* and nematodes presented evidence of the role of growth-factor-like substances in the selection of the direction of differentiation.

Furthermore, mesoderm differentiation in *Xenopus* has been shown to be regulated by growth factors. Although growth factors are soluble materials, cell surface molecules are involved in the regulation of the differentiation by growth factors and related molecules: some growth-factor-like molecules are cell surface molecules, and growth factors exert their effects mostly by binding to cell surface receptors.

4.1 GROWTH FACTORS IN HAEMATOPOIESIS

As mentioned in the introduction, many growth factors and related molecules play essential roles in haematopoiesis. In addition to erythropoietin and CSFs, some interleukins need special attention. IL-3 produced by activated T cells is a potent stimulator of haematopoiesis; it stimulates the initial proliferation and maturation of myeloid stem cells (CFU-S), progenitor cells of granulocyte/macrophage lineage (GM-CFS), mast cell precursors and pre T cells [16, 17, 18]. IL-4, 5 and 6, all of which are secreted by activated T cells, promote the maturation of antigen-stimulated B cells into plasma cells (Table 4.1). Thus, these factors form a basis of T cell–B cell cooperation in antibody production. As mentioned before, stimulation of B cells is a function of IL-4–6. IL-4 stimulates T cells, mast cells [19] and can enhance or antagonize factor-dependent growth of haematopoietic progenitor cells [20].

Growth factors can regulate the direction of differentiation in haematopoiesis, by selectively stimulating progenitor cells leading to a specific cell lineage. It is possible that a haematopoietic cell yields two different daughter cells according to a programmed schedule [21] and growth factors in the environment determine whether one, two or no daughter cells survive and proceed with the scheduled cell lineage. This mode of growth factor action can be called the 'selective growth model', and can be applied to various aspects of haematopoiesis and other differentiation systems (Figure 4.2).

Growth factors can regulate the direction of differentiation in a more direct way. In the presence of a growth factor (or at a given concentration of a growth factor), a haematopoietic cell may yield daughter cells of a given property and in the presence of another growth factor (or at a different concentration of the first growth factor) the same cell may yield daughter cells of a different property. This model of growth factor action may be called the 'selective differentiation model' (Figure 4.2). Without precise analysis of the differentiation system, it is difficult to determine by what mechanism a growth factor can influence the direction of differentiation. However, there are increasing amounts of evidence to indicate that the 'selective differentiation model' is applicable at least to certain

I Selective growth model

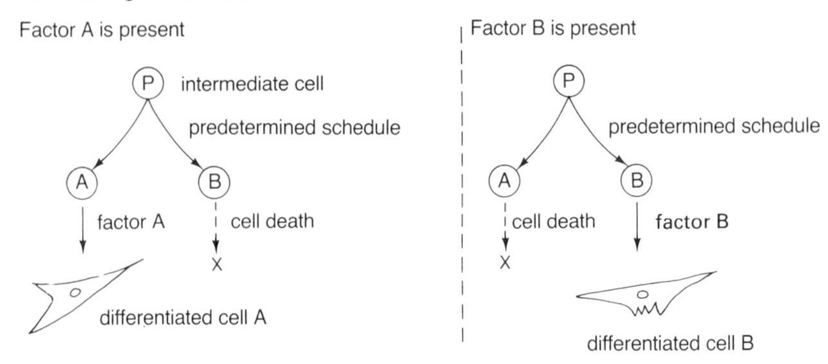

II Selective differentiation model

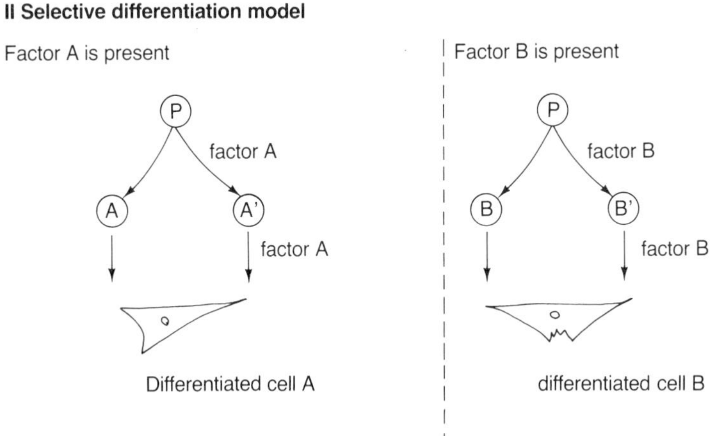

Figure 4.2 A simplified version of two models on regulation of differentiation by growth factors. (A), (A'), (B) and (B'), cells are the first progenies of (P) cell.

differentiation systems. The following is a classic example:

Bone marrow cells differentiate to granulocytes and monocytes in the presence of GM-CSF. Metcalf found that at relatively high concentrations of the factor, cells differentiated more toward granulocytes, and at lower concentrations toward macrophages [22]. This finding can be interpreted in several ways:

1. From common precursor cells, macrophages differentiate in low GM-CSF, while in high GM-CSF granulocytes are formed.

2. In addition to common precursor cells leading to both cell types, there are precursor cells yielding only macrophages and those yielding only granulocytes. In lower concentrations of the factor, those yielding only macrophages are stimulated, while in higher concentrations of the factor, those yielding only granulocytes are stimulated.

In order to obtain clearer insight into the effects of growth factor concentration, Metcalf carried out the following experiment [22]:

Bone marrow cells were cultured with GM-CSF. When the cells divided once, progeny cells were picked up by a micropipet and one was cultured in high GM-CSF and the other in low GM-CSF. As a result, the presence of three different cell types became clear: cells which differentiated to macrophages in low GM-CSF and to granulocytes in high GM-CSF; cells which always differentiated to macrophages; and cells which always differentiated to granulocytes. As above, a cell population changes its direction of differentiation responding to the concentration of a growth factor. The precursor cells divide more rapidly in higher concentrations of GM-CSF. Therefore, it is interesting to speculate that the bipotential cells differentiate to macrophages when the growth rate is low, and to granulocytes when the growth rate is high.

4.2 GLIA CELL DIFFERENTIATION

Growth factors are also important in the regulation of glia cell differentiation. Glia cells are present in the brain and the spinal cord and assist the function of nerve cells. Glia cells include astrocytes, which have projections touching neurite or blood vessels and transport nutrients to the nerve, and oligodendrocytes, which produce the myelin sheath which surrounds the neurites.

Raff studied glia cell differentiation in optic nerve cells of the rat embryo. Astrocytes, oligodendrocytes and precursor cells were recovered by isolating the neurite of the optic nerve, since these cells attach to the neurites. Immunochemical markers disclosed the presence of two kinds of astrocytes in the optic nerve: one not expressing the carbohydrate antigen, A2B5 (astrocyte I) and another expressing the antigen (astrocyte II). Oligodendrocytes and the precursor cells could be also identified by immunochemical markers (Table 4.2). Raff demonstrated that the precursor cells differentiated to oligodendrocytes in the absence of fetal calf serum and to astrocyte II in its presence. Therefore, a factor present in fetal calf serum determines the fate of the precursor cell [23].

The time schedule of the appearance of oligodendrocyte was subsequently investigated. When cells attached to the neurite of 17th day embryos were cultured, oligodendrocytes appeared after 2 days. This is 2 days earlier than

Table 4.2 Distribution of markers in the glia cell lineage

	Astrocyte I	Astrocyte II	Oligo-dendrocyte	Progenitor cell
Glia acidic protein	+	+	–	–
Galactocerebroside	–	–	+	–
A2B5	–	+	+ → –	+

the natural appearance of oligodendrocytes during rat embryogenesis. The natural schedule of the appearance of oligodendrocytes can be recovered by culturing the cells on a monolayer of astrocyte I; the precursor cells differentiate to oligodendrocytes just as scheduled in the embryo, namely after 4 days [24]. The result can be interpreted as follows (Figure 4.3). In the absence of a serum factor which induces the differentiation to astrocyte II, the precursor cells differentiate to oligodendrocytes when they stop cell division. Without a factor secreted by astrocyte I, the precursor cells divide only for a short time period and differentiate to oligodendrocytes after 2 days in culture. With a factor provided by astrocyte I, the precursor cells are encouraged to divide; cells become unresponsive to the factor within 4 days, stop division and differentiate to oligodendrocytes (Figure 4.3). The factor secreted by type I astrocyte has been identified as PDGF [25, 26].

4.3 GROWTH FACTORS IN EARLY MAMMALIAN EMBRYOGENESIS

Without doubt, growth factors are also important in early mammalian embryogenesis. This has been shown by Heath and co-workers using an *in vitro* differentiation system of EC cells [27]. EC cells secrete a growth factor, embryonal carcinoma cell-derived growth factor (ECDGF), which promotes the growth of many cells, including END cells. The END cell is a differentiated derivative of EC cells and resembles extraembryonic mesoderm cells. On the other hand, END cells secrete a factor resembling insulin-like growth factor-II, which promotes the growth of EC cells. Therefore, the subpopulations of early embryonic cells appear to interact to each other, partly by growth factors. Furthermore, EC cells do not express functional receptors for insulin and EGF, both of which come to be expressed after *in vitro* differentiation. Furthermore, *int-2* proto-oncogene, a member of FGF gene family, comes to be expressed after *in vitro* differentiation of EC cells (section 4.7).

These observations imply a complicated mode of regulation and function of the growth factor systems in early mammalian embryogenesis.

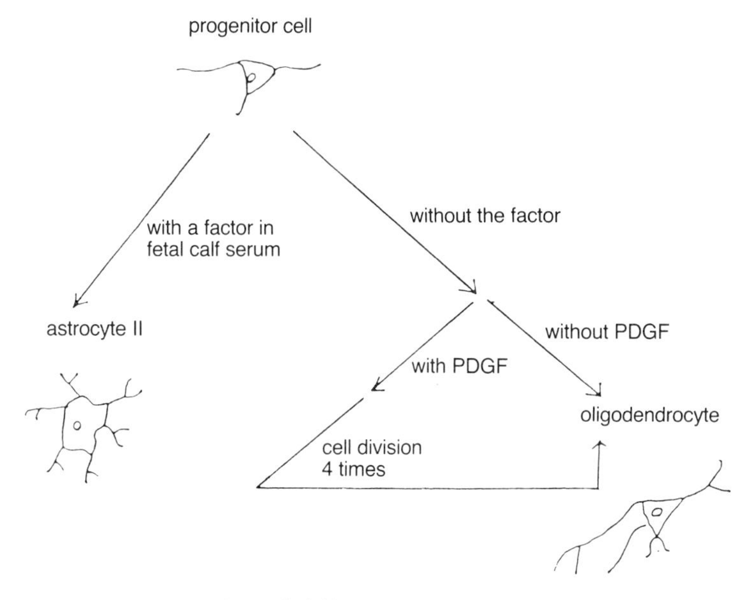

Figure 4.3 Cell fates in glia cell differentiation.

4.4 EGF-LIKE REPEATS IN NEUROGENESIS

Molecular genetic analysis of a *Drosophila* developmental mutant, *Notch*, provided strong evidence for the role of growth-factor-related cell surface molecules in the determination of cell fate. *Notch* is a locus active at several developmental stages. A homozygous embryo with a null *Notch* mutation (*N* or *Notch*) is embryonic lethal. Embryos heterozygous in the *Notch* mutation survive, but the adult has abnormalities in hydrodermis-derived tissues. A prominent abnormality is a V-shaped notch in the wing [28]. The principal defect of *N* lies in the modulation of the developmental fate of ectoderm cells. The neurogenic region, which is a part of the ectoderm, consists of about 1800 cells. Part of them (about 400 cells) differentiate along the neuronal pathway, migrate to the interior of the embryo, and form nervous tissue. The other cells remain at the surface of the embryo and become hypodermal cells. Neuroblasts (a precursor to nerve cells) probably exert inhibitory effects on uncommitted cells differentiating to neuroblasts. In the *N* embryos, this modulation does not occur properly; more cells become nerve cells in *N* embryos and the nervous tissue contains 3 times as many cells as normal. On the other hand, hypoderm develops poorly in the *N* embryo; this poor development results in the death of the *N* embryo later in embryogenesis.

Artavanis-Tsakonas *et al.* cloned the gene of *Notch* locus and, from the cDNA sequence, predicted the amino acid sequence of *Notch* protein [29, 30]. The predicted protein with 2703 amino acids, is probably a transmembrane protein, since it contains both a transmembrane domain and a signal sequence (Figure 4.4). The putative extracellular domain has 36 repeated structures of EGF-like repeats (Figure 4.5). The characteristic arrangement of cysteine residues in the EGF active site is well conserved in the repeat. Probably, an EGF-like homology unit in *Notch* protein has a disulphide bridge and three-dimensional structure, as seen in EGF. An important question is whether the *Notch* protein exerts a function in a membrane-bound form or serves as a precursor of soluble factors. The former possibility is favoured, since in *N* and wild type mosaic flies, cells must synthesize *Notch* proteins by themselves in order to behave normally. EGF-like repeats are also found in laminin, blood clotting factors and LDL receptor. An EGF-like motif is probably favoured if the molecule is to be recognized strictly by other proteins. *Notch* protein is thought to be recognized by a cell surface molecule in the neighbouring cell and thereby regulate the direction of differentiation. EGF, a soluble growth factor, might have evolved from an intercellular-communicating protein with EGF-like repeats such as *Notch*.

In addition to *Notch* mutation, several other mutations are known to be associated with *Notch* locus. For example, *Abruptex* (*Ax*) mutations are other dominant mutations in the locus, and cause a gapping of the wing veins. *Ax* mutants are classified into three groups, Ax^{E} (enhancer), Ax^{S} (suppressor) and Ax^{L} (lethal). Ax^{E} enhances the *N* phenotype of the nicked wing in heterozygotes (N/Ax^{E}), while Ax^{S} suppresses the *N* phenotype in heterozygotes (N/Ax^{S}). Ax^{S} and Ax^{E} are homozygous viable, while Ax^{S}/Ax^{E} heterozygotes are lethal. Ax^{L} mutants are homozygous lethal. A sequence analysis of *Notch* locus has been performed for 7 *Ax* mutants [31]. Every mutation is caused by a single amino acid change in one of 5 EGF-like repeats present at the C-terminal side of the 36 EGF-like repeats. Ax^{L} mutants are caused by a change in a cysteine residue (Figure 4.5). While no

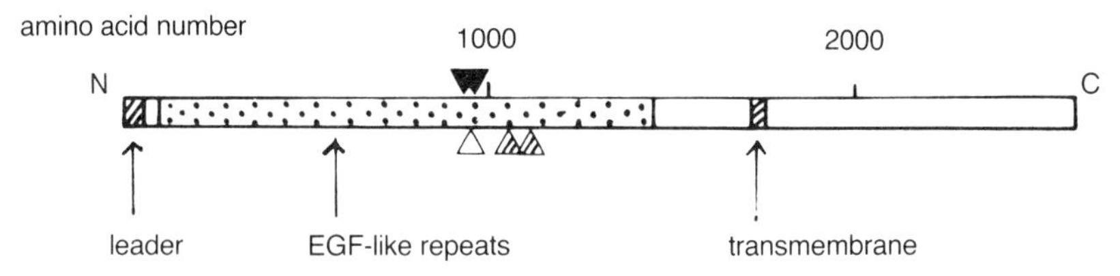

Figure 4.4 Model of the *Notch* gene product. The following indicate the locations of Ax mutations: ▼: suppressor; △: lethal; ◬: enhancer. (Based on [30] and [31].)

Notch	C - - - X S X P C X N G G T C - - - X D X X X X F/Y X C X C X X G F/Y X G X X C
Factor IX precursor	C - - - E S N P C L N G G S C - - - K D D I N S Y E C W C P F G F E G K N C
Plasminogen activator	C - - - S E P R C F N G G T C Q Q A L Y F S D F V - C Q C P E G F A G K C C
EGF	C P S S Y D G Y C L N G G V C - M H I E S L D S Y T C N C V I G Y S G D R C

Figure 4.5 Homology of *Notch* repeats with EGF and related molecules. The top panel is the consensus sequence of *Notch* repeats. X: any amino acid. (Taken from [30].)

two of the 35 EGF-like repeats of *Notch* protein are identical, the bias of amino acid mutation in the C terminal side in *Ax* mutants indicates that EGF-like repeats in the N-terminal side and the C-terminal side have different functions. The lethality of Ax^S/Ax^E may mean that *Notch* polypeptide interacts with itself and that the two mutated polypeptides cannot make functional interactions.

Delta (*Dl*) is another neurogenic gene of *Drosophila*, and is also involved in the determination of an epidermal or neural fate. *Dl* protein is again thought to be a transmembrane protein with 9 EGF-like repeats [32].

The EGF-like repeated structure is also involved in determination of cell fate in *C. elegans*. During embryogenesis, precursors to vulvis can differentiate to three different sets of cells, and this decision can be changed by external factors. The *lin-12* gene product is involved in this decision process. Greenwald and co-workers cloned *lin-12* gene and determined the structure. *lin-12* protein has overall similarity to *Notch* protein and has 13 EGF-like repeats [33, 34].

4.5 MESODERM INDUCTION

Striking information on the role of growth factors in the determination of cell fate came from an analysis of the early embryogenesis of *Xenopus* embryos. As described in Chapter 1, the vegetal portion of the early embryo develops spontaneously into endoderm and dictates the development of the neighbouring ectoderm cells into mesoderm cells.

One factor isolated from chick embryo, and another partially purified from *Xenopus* cells, can induce mesoderm formation. Slack *et al.* found that a basic fibroblast growth factor (FGF)-like material plays an important role in the mesoderm-inducing activity of endoderm [35]. They ablated the ectoderm of *Xenopus* embryos and cultured the explant. The effects of various growth factors on the developmental fate of the explant have been examined. Without added material, the explant rounds up and forms epidermis around the outer surface, and no mesoderm cells differentiate. In the presence of basic FGF, the cell aggregates are elongated in shape and swollen by fluid uptake (Figure 4.6). A histological examination reveals clear signs of the development of the mesoderm. Other growth factors tested show no mesoderm inducing activity. Kimelman and Kirschner have confirmed the finding of Slack *et al.* and furthermore, have found that TGF-β enhances the action of FGF [36]. TGF-β alone has no such effect. TGF-β2 is a factor with strong homology to TGF-β (which is alternatively called TGF-β1). Rosa *et al.* have found that TGF-β2 alone induces mesoderm formation [37]. Furthermore, antibodies to TGF-β2 but not TGF-β1 inhibit the process. From these results, there is no doubt that growth factors play important roles in mesoderm induction. The remaining question is to identify the nature of the growth factors actually working in *Xenopus* embryos.

A cDNA clone specifying a protein sequence similar to FGF has been isolated from *Xenopus* embryos. Furthermore, Weeks and Melton have found that Vg1, a maternal mRNA localized in the vegetal hemisphere of *Xenopus* embryos, specifies a protein showing 38% homology to TGF-β [38]. Thus, an FGF and TGF-like molecule appears certainly to be involved in the induction of mesoderm. The localization of Vg1 mRNA is also of special interest. In oocytes, the RNA is found in the cortical shell at the vegetal pole. In the unfertilized egg, the RNA spreads toward the animal pole and forms a broader band, although it is still restricted to the vegetal

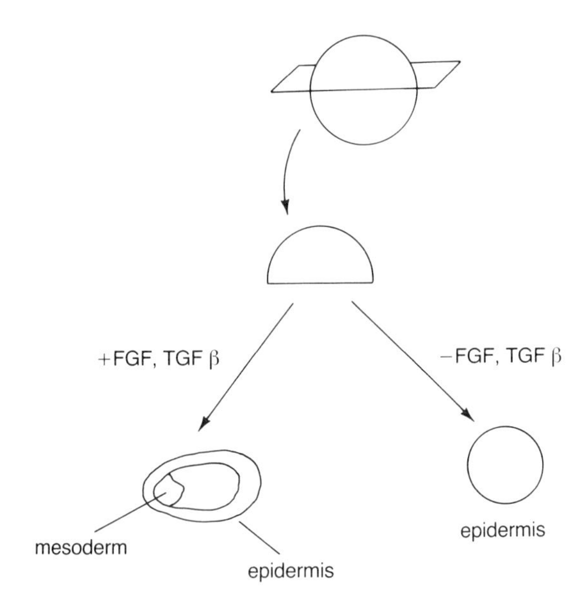

Figure 4.6 Requirement of growth factors for mesoderm induction in Xenopus embryos.

hemisphere. During cleavage, the RNA is distributed to most of the vegetal cells. Thus, the Vg1 molecule may not only play an important role in mesoderm induction but may also be important in defining the endoderm fate of vegetal cells.

4.6 TGF-β AND THE SUPERFAMILY IN OTHER DEVELOPMENTAL SYSTEMS

The distribution of TGF-β between the 11th and 18th day of gestation during mouse embryogenesis has been examined [39]. It is present in mesenchyme, mesenchyme-derived tissues (connective tissue, cartilage and bone) and neural crest mesenchyme-derived tissues (palate, larynx, facial mesenchyme, nasal sinuses, meninges and teeth). Intense expression is found where mesenchyme interacts with adjacent epithelium, and mesenchyme and mesoderm undergo remodelling. TGF-β is known to enhance the synthesis of matrix components such as collagen and fibronectin and fibronectin receptor (integrin) [40, 41]. Thus, it is likely that TGF-β participates in the regulation of organogenesis by altering the metabolism of cell surface components.

TGF-β plays a decisive role in chondrogenesis. Demineralized bone was found to have a factor which induced cartilage formation upon *in vivo* administration [42]. The action of the factor can be measured *in vitro*, since it promotes the differentiation of muscle mesenchymal cells to chondroblasts. The factor, subsequently purified, has turned out to be indistinguishable from TGF-β [43]. Thus, is is concluded that TGF-β promotes the differentiation of muscle mesenchymal cells into chondroblasts.

TGF-β superfamily also exists in *Drosophila* and is critical in determining cell fate. The decapentaplegic gene complex (DPP-C) regulates several events in pattern formation in *Drosophila* development. Three genetic regions have been identified in this gene complex: shortvein (*shv*), Haplo insufficiency near decapentaplegic (*Hin-d*) and decapentaplegic region (*dpp*). A homozygous *Hin-d* null mutant develops into a completely ventralized embryo and is embryonic lethal. Mutations of *dpp* region result in the loss of adult epidermal structures derived from the centres of the imaginal disks. *Shv* mutants have defects in venation of the wing and in the maxillary palps of the head. Gelbart and co-workers have determined the structure of cDNA corresponding to *Hin-d* region [44]. The predicted protein (66 K) has a leader sequence, and about 100 amino acid residues in the C terminal region have 36% homology with TGF-β (Figure 4.7). Cysteine residues preserved in TGF-β superfamily are also conserved in this protein.

Besides Vg1 and DPP-C, three polypeptide factors are related to TGF-β; they are inhibins, activins and Müllerian inhibitory substance (MIS) (Figure

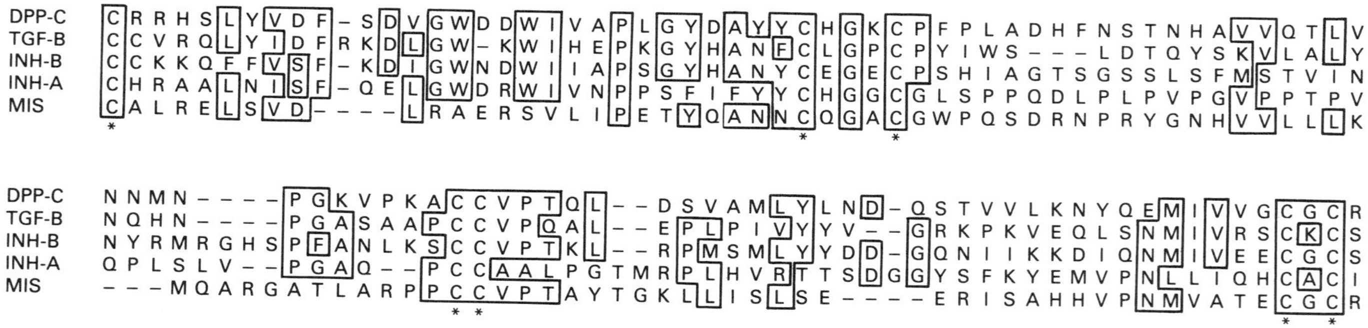

```
DPP-C   C R R H S L Y V D F - S D V G W D D W I V A P L G Y D A Y Y C H G K C P F P L A D H F N S T N H A V V Q T L V
TGF-B   C C V R Q L Y I D F R K D L G W - K W I H E P K G Y H A N F C L G P C P Y I W S - - - L D T Q Y S K V L A L Y
INH-B   C C K K Q F F V S F - K D I G W N D W I I A P S G Y H A N Y C E G E C P S H I A G T S G S S L S F M S T V I N
INH-A   C H R A A L N I S F - Q E L G W D R W I V N P P S F I F Y Y C H G G C G L S P P Q D L P L P V P G V P P T P V
MIS     C A L R E L S V D - - - - L R A E R S V L I P E T Y Q A N N C Q G A C G W P Q S D R N P R Y G N H V V L L L K
        *                                                             *       *

DPP-C   N N M N - - - - P G K V P K A C C V P T Q L - - D S V A M L Y L N D - Q S T V V L K N Y Q E M I V V G C G C R
TGF-B   N Q H N - - - - P G A S A A P C C V P Q A L - - E P L P I V Y Y V - - G R K P K V E Q L S N M I V R S C K C S
INH-B   N Y R M R G H S P F A N L K S C C V P T K L - - R P M S M L Y Y D D - G Q N I I K K D I Q N M I V E E C G C S
INH-A   Q P L S L V - - P G A Q - - P C C A A L P G T M R P L H V R T T S D G G Y S F K Y E M V P N L L I Q H C A C I
MIS     - - - M Q A R G A T L A R P P C C V P T A Y T G K L L I S L S E - - - - E R I S A H H V P N M V A T E C G C R
                                   * *                                                                 *   *
```

Figure 4.7 Conserved structural feature of TGF-β family. Sequences of C-terminal side are shown. Asterisks are conserved cysteines. DPP-C: *Hin-d* region of DPP-C; INH: inhibin; MIS: Müllerian inhibitory substance. (Cited from [44].) © 1987. Macmillan Magazines Ltd.

4.7) [45]. Inhibins (inhibin A and inhibin B) are present in ovarian and testicular fluid and suppress the production of follicle-stimulating hormone (FSH). On the other hand, activins found in the same sources as inhibins promote the production of FSH. Inhibins and activins are composed of subunits (Figure 4.8). A subunit of inhibin A (βA) is a subunit of activin A and AB. A subunit of inhibin B is also a subunit of activin AB. A protein factor that stimulates the differentiation of Friend leukemia cells and K-562 cells has been found to be identical to βA [46]. MIS is a glycoprotein secreted by the Sertoli cells of the testes and causes the regression of the Müllerian duct, which in the normal female embryo develops into the uterus, vagina and fallopian tubes. Thus, MIS is important in the development of the male genital organ; in addition to testosterone which is responsible for the differentiation of the epididymis, vas deferens and seminal vesicles from the Wolffian duct. MIS (140 K) is composed of two subunits of 74 and 70 K. The C terminal domain of MIS is homologous to TGF-β [47]. As above, TGF-β family comprises several polypeptides with different functions and is expected to include more species of developmental significance.

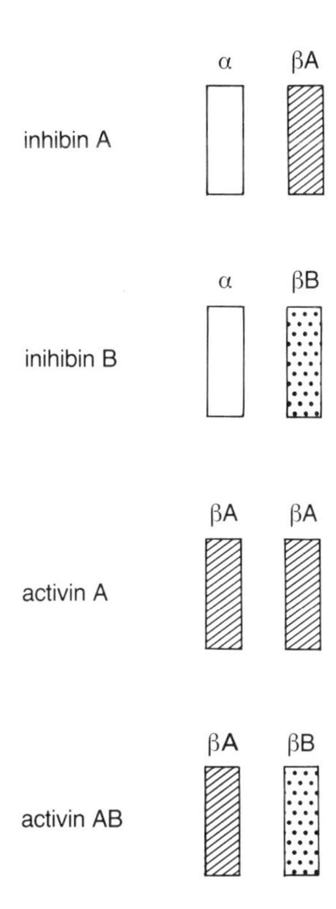

Figure 4.8 Subunit composition of inhibins and activins.

4.7 *int-1* AND *int-2*

Still another growth-factor-like material has been implicated in cell differentiation. *int-1* and *int-2* have been found as photo-oncogenes present in the insertion site of mouse mammary tumour virus. Insertional activation of these genes is expected to play an important role in the development of mammary tumours. As mentioned earlier *int-2* is related to fibroblast growth factor (FGF); *int-1* is not related to other proteins, however the presence of a leader sequence and high content of cysteine suggest that *int-1* is also a growth factor. The *Drosophila* compartment of *int-1* (*Dint-1*) has been found to be identical to *wingless* gene [48]. *Wingless* is a segment polarity gene; in a lethal mutant of *wingless* locus, the posterior part of each segment is replaced by an apparent mirror-image duplication of the anterior part. *Dint-1* is expressed in a segmental pattern in the embryo (Figure 4.9).

In the adult mouse, *int-1* expression is detected only in postmeiotic germ cells [49]. In embryos, *int-1* RNA is detected only in specific regions of the neural plates and its derivatives only between 9 and 14½ days of development [50]. *int-1* protein is probably involved in a step of neurogenesis. *int-2* is expressed during *in vitro* differentiation of EC cells. On the other hand, another member of FGF family, *hst* (an oncogene found in human stomach cancer), is expressed in EC cells and becomes unexpressed during the differentiation process [51]. Thus, the two growth factor-like substances of the same family, are probably regulated in a reciprocal way during early mouse embryogenesis.

4.8 A PUTATIVE RECEPTOR WITH TYROSINE KINASE ACTIVITY IN PHOTORECEPTOR DIFFERENTIATION

The critical role of growth factors (and the related membrane proteins) in determining cell fate implies that the receptors for the factors are equally important in the process. As described in Chapter 2, the receptors so far

Figure 4.9 Localization of *Dint-1* (*wingless*) mRNA revealed by *in situ* hybridization in a *Drosophila* embryo. (Reprinted from [48].) © Cell Press.

disclosed are transmembrane proteins. The importance of the receptors is deduced from the observations that, in mesoderm induction in *Xenopus* embryogenesis, ectoderm is responsive to the growth factors only for restricted periods; most probably only during the period receptors for the factors are expressed on the surface.

The critical importance of the receptors has been illustrated again by the developmental genetics of *Drosophila*. The compound eye of the fly consists of about 800 ommatidia (unit eye). Each ommatidia contains 8 photoreceptor cells and 12 other cells. The photoreceptor cells can be divided into three classes (R1–R6, R7 and R8). Of these, R7 is the only photoreceptor dedicated to the reception of ultraviolet light. Therefore, mutant strains defective in ultraviolet light perception include mutants affected in development of the R7 photoreceptor. Several such mutants have been found to be caused by mutations of a single genetic locus, *sevenless* (*sev*). R7 cells are absent in the mutants (Figure 4.10); differentiation of a precursor cell to R7 appears to be affected by the mutation, and the precursor cell is believed to differentiate into a non-neuronal cell.

sev gene has been cloned and the sequence determined. *sev* gene encodes a transmembrane protein whose cytoplasmic domain has extensive homology with tyrosine kinase, such as those found in EGF receptor, *src* and *ros* oncogene. Therefore, *sev* gene appears to specify a receptor protein analogous to EGF receptor [51]. It is likely that a membrane-bound growth-factor-like structure in the neighbouring cell, as found in the gene product of *Notch*, interacts with *sev* gene product and gives the positional information

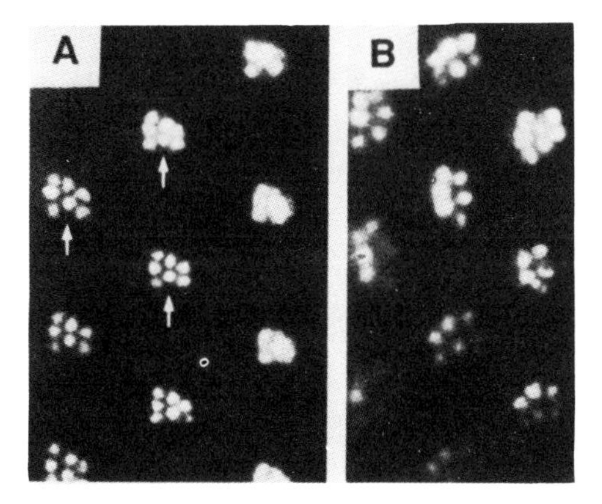

Figure 4.10 Comparison of rhabdomere pattern of (A) a wild-type fly and (B) a *sev* fly. White arrows indicate R7 cells, which are present in a wild-type fly but not in a *sev* fly. (Reprinted from [52].) © 1987. AAAS.

to yield R7 cell. Since *sev* gene product is expressed not only in the R7 precursor cell but also in other cells, another factor(s) must be required to specify the R7 phenotype [53, 54].

4.9 *c-kit* PROTO-ONCOGENE AND *W* LOCUS

Recent results also indicate that *c-kit* proto-oncogene, encoding a putative transmembrane receptor, is the dominant-white spotting (*W*) locus of the mouse [55, 56]. *c-kit* polypeptide is a transmembrane one, has a cytoplasmic tyrosine kinase domain and shows homology with receptors for CSF-1 and PDGF. In most cases, homozygotes with mutations in *W* locus are sterile, have extensive white-spotting and a severe anaemia. *W* locus appears to be required for the proliferation, migration or differentiation of three cell populations, namely haematopoietic stem cells, migrating melanoblasts and primordial germ cells. Chromosome mapping studies have indicated the possibility that *c-kit* and *W* locus may be identical. Close analysis of *W* mutants verified that this is indeed the case. Although the ligand of *c-kit* gene product is not known, most probably it is a growth factor. Therefore, the factor and the receptor (*c-kit*) appear to be important in the three systems of differentiating cells mentioned above.

REFERENCES

1. Levi-Montalcini, R. (1987) *EMBO J.*, **6**, 1145–54
2. Savage, C.R. Jr., Hash, J.H. and Cohen, S. (1973) *J. Biol. Chem.*, **248**, 7669–73
3. Golde, D.W. and Casson, J.C. (1988) *Sci. Amer.*, **259** (Jul), 34–42
4. Metcalf, D. (1985) *Science*, **229**, 16–22
5. Clark, S.C. and Kamen, R. (1987) *Science*, **236**, 1229–37
6. Miyake, T., Kung, C.K. and Goldwasser, E. (1977) *J. Biol. Chem.*, **252**, 5558–64
7. Burgess, A.W., Camakaris, J. and Metcalf, D. (1977) *J. Biol. Chem.*, **252**, 1998–2003
8. Campbell, H.D., Tucker, W.Q.J., Hort, Y., Martinson, M.E., Mayo, G. *et al.* (1987) *Proc. Natl. Acad. Sci.*, **84**, 6629–33
9. Kinashi, T., Harada, N., Severinson, E., Tanabe, T., Sideras, P. *et al.* (1986) *Nature*, **324**, 70–3
10. De Larco, J.E. and Todaro, G.J. (1978) *Proc. Natl. Acad. Sci.*, **75**, 4001–5
11. Sporn, M.B., Roberts, A.B., Wakefield, L.M. and Assolan, R.K. (1986) *Science*, **233**, 532–4
12. Derynck, R., Jarrett, J.A., Chen, E.Y., Eaton, D.H., Bell, J.R. *et al.* (1985) *Nature*, **316**, 701–5
13. Doolittle, R.F., Hunkapiller, M.W., Hood, L., Devare, S., Robbins, K. *et al.* (1983) *Science*, **221**, 275–7
14. Peters, G., Brookes, S., Smith, R. and Dickson, C. (1983) *Cell*, **33**, 369–77
15. Taira, M., Yoshida, T., Miyagawa, K., Sakamoto, H., Terada, M. and Sugimura, T. (1987) *Proc. Natl. Acad. Sci.*, **84**, 2980–4
16. Burgess, A.W. (1985) *Ciba Foundation Symp.*, **116**, 148–68

17. Dexter, T.M., Heyworth, C.M. and Whetton, A.D. (1985) *Ciba Foundation Symp.*, **116**, 129–47
18. Palacios, R., Henson, G., Steinmetz, M. and McKearn, J.P. (1984) *Nature*, **309**, 126–31
19. Lee, F., Yokota, T., Otsuka, T., Meyerson, P., Villaret, D. *et al.* (1986) *Proc. Natl. Acad. Sci.*, **83**, 2061–5
20. Rennick, D., Yang, G., Muller-Sieburg, C., Smith, C., Arai, N. *et al.* (1987) *Proc. Natl. Acad. Sci.*, **84**, 6889–93
21. Suda, T., Suda, J. and Ogawa, M. (1984) *Proc. Natl. Acad. Sci.*, **81**, 2520–4
22. Metcalf, D. (1980) *Proc. Natl. Acad. Sci.*, **77**, 5327–30
23. Raff, M.C., Miller, R.H. and Noble, M. (1983) *Nature*, **303**, 390–6
24. Raff, M.C., Abney, E.R. and Fok-Seang, J. (1985) *Cell*, **42**, 61–9
25. Richardson, W.D., Pringle, N., Mosley, M.J., Westermark, B. and Dubois-Dalcq, M. (1988) *Cell*, **53**, 309–19
26. Noble, M., Murray, K., Stroobant, P., Waterfield, M.D. and Riddle, P. (1988) *Nature*, **333**, 560–2
27. Heath, J.K. and Rees, A.R. (1985) *Ciba Foundation Symp.*, **116**, 3–22
28. Kidd, S., Lockett, T.J. and Young, M.W. (1983) *Cell*, **34**, 421–33
29. Artavanis-Tsakonas, S., Muskavitch, M.A.T. and Yedvobnick, B. (1983) *Proc. Natl. Acad. Sci.*, **80**, 1977–81
30. Wharton, K.A., Johansen, K.M., Xu, T. and Artavanis-Tsakonas, S. (1985) *Cell*, **43**, 567–81
31. Kelley, M.R., Kidd, S., Deutsch, W.A. and Young, M.W. (1987) *Cell*, **51**, 539–48
32. Vässin, H., Bremer, K.A., Knust, E. and Campos-Ortega, J.A. (1987) *EMBO J.*, **6**, 3431–40
33. Greenwald, I. (1985) *Cell*, **43**, 583–90
34. Yochem, J., Weston, K. and Greenwald, I. (1988) *Nature*, **335**, 547–54
35. Slack, J.M.W., Darlington, B.G., Heath, J.K. and Godsave, S.F. (1987) *Nature*, **326**, 197–200
36. Kimelman, D. and Kirschner, M. (1987) *Cell*, **51**, 864–77
37. Rosa, F., Roberts, A.B., Danielpour, D., Dart, L.L., Sporn, M.B. and Dawid, I.B. (1988) *Science*, **239**, 783–5
38. Weeks, D.L. and Melton, D.A. (1987) *Cell*, **51**, 861–7
39. Heine, U.I., Munoz, E.F., Flanders, K.C., Ellingsworth, L.R., Lam, H.P. *et al.* (1987) *J. Cell. Biol.*, **105**, 2861–76
40. Ignotz, R.A., Endo, T. and Massagué, J. (1987) *J. Biol. Chem.*, **262**, 6443–6
41. Ignotz, R.A. and Massagué, J. (1987) *Cell*, **51**, 189–97
42. Urist, M.R. (1965) *Science*, **150**, 893–9
43. Seyedin, S.M., Thompson, A.Y., Bentz, H., Rosen, D.M., McPherson, J.M. *et al.* (1986) *J. Biol. Chem.*, **261**, 5693–5
44. Padgett, R.W., Johnston, R.D.S. and Gelbart, W.M. (1987) *Nature*, **325**, 81–4
45. Massagué, J. (1987) *Cell*, **49**, 437–8
46. Murata, M., Eto, Y., Shibai, H., Sakai, M. and Muramatsu, M. (1988) *Proc. Natl. Acad. Sci.*, **85**, 2434–8
47. Cate, R.L., Mattaliano, R.J., Hession, C., Tizard, R., Farber, N.M. *et al.* (1986) *Cell*, **45**, 685–98
48. Rijsewijk, F., Schuermann, M., Wagenaar, E., Parren, P., Wiegel, D. and Nusse, R. (1987) *Cell*, **50**, 649–57
49. Shackleford, G.M. and Varmus, H.E. (1987) *Cell*, **50**, 89–95
50. Wilkinson, D.G., Bailes, J.A. and McMahon, A.P. (1987) *Cell*, **50**, 79–88

51. Yoshida, T., Muramatsu, H., Muramatsu, T., Sakamoto, H., Katoh, O. *et al.* (1988) *Biochem. Biophys. Res. Commun.*, **157**, 618–25
52. Hafen, E., Basler, K., Edstroem, J.-E. and Rubin, G.M. (1987) *Science*, **236**, 55–63
53. Tomlinson, A., Bowtell, D.D.L., Hafen, E. and Rubin, G.M. (1987) *Cell*, **51**, 143–50
54. Banerjee, U., Renfranz, P.J., Hinton, D.R., Rabin, B.A. and Benzer, S. (1987) *Cell*, **51**, 151–8
55. Geissler, E.N., Ryan, M.A. and Housman, D.E. (1988) *Cell*, **55**, 185–92
56. Chabot, B., Stephenson, D.A., Chapman, V.M., Besmer, P. and Bernstein, A. (1988) *Nature*, **335**, 88–9
57. Angeletti, R.H. and Bradshaw, R.A. (1971) *Proc. Natl. Acad. Sci.*, **68**, 2417–20
58. Gspodarowics, D. (1987) *Methods Enzymol.*, **147**, 106–19
59. Raines, E.W. and Ross, R. (1982) *J. Biol. Chem.*, **257**, 5154–60
60. Gough, N.M., Gough, J., Metcalf, D., Kelso, A., Grail, D. *et al.* (1984) *Nature*, **309**, 763–7
61. Lomedico, P.T., Gubler, U., Hellmann, C.P., Dukovich, M., Giri, J.G. *et al.* (1984) *Nature*, **312**, 458–62
62. Taniguchi, T., Matsui, H., Fujita, T., Takaoka, C., Kashima, N. *et al.* (1983) *Nature*, **302**, 305–10
63. Fung, M.C., Hapel, A.J., Ymer, S., Cohen, D.R., Johnson, R.M. *et al.* (1984) *Nature*, **307**, 233–7
64. Yokota, T., Lee, F., Rennick, D., Hall, C., Arai, N. *et al.* (1984) *Proc. Natl. Acad. Sci.*, **81**, 1070–4
65. Noma, Y., Sideras, P., Naito, T., Bergstedt-Lindquist, S., Azuma, C. *et al.* (1986) *Nature*, **319**, 640–6
66. Hirano, T., Yasukawa, K., Harada, H., Taga, T., Watanabe, Y. *et al.* (1986) *Nature*, **324**, 73–6
67. Zilberstein, A., Ruggieri, R., Korn, J.H. and Revel, M. (1986) *EMBO J.*, **5**, 2529–37
68. May, L.T., Helfgott, D.C. and Sehgal, P.B. (1986) *Proc. Natl. Acad. Sci.*, **83**, 8957–61

5 Cell adhesion molecules

Adhesion of cells to neighbouring cells and to the extracellular matrix is of crucial importance in embryogenesis. During embryogenesis, a type of cell adheres to a cell of the same type or to a cell of another type. When required, cells detach from a cell group, migrate and stop at a defined point. A variety of cell to cell and cell to matrix adhesion proteins are involved in the process. Furthermore, some of the extracellular information regulating differentiation appears to be transmitted into the cell by cell adhesion molecules.

In 1955, Townes and Holtfreter [1] dissociated amphibian embryos in alkaline solution. When cells from different regions were mixed and allowed to form cell aggregates, each cell type usually sorted out and segregated. For example, when mesoderm cells and endoderm cells were mixed, endoderm cells adhered to each other and were located in the central portion of cell aggregates, while the mesoderm cells covered the cluster of endoderm cells. Moscona dissociated chick embryonic cells by trypsin; the dissociated cells were gently swirled and were allowed to sort. When retina cells and liver cells were co-aggregated, retina cells sorted into the inner portion of the aggregate [2]. Since then, molecular entities responsible for selective cell adhesion have been investigated by many researchers, and various cell surface molecules involved in cell adhesion have been identified and isolated. Recombinant DNA technology has allowed the elucidation of the protein structures of the adhesion molecules. Transmembrane proteins involved in cell adhesion are largely classified into three groups: the immunoglobulin superfamily, the cadherin family and the integrin superfamily.

5.1 IMMUNOGLOBULIN SUPERFAMILY

5.1.1 N-CAM

Increasing numbers of adhesive molecules are being assigned to the immunoglobulin superfamily. The prototype of the adhesion molecule of this class is N-CAM (neural cell adhesion molecule). N-CAM is involved in the cell to cell adhesion of various cells, notably neural cells. In the middle of the 1970s, Edelman and co-workers [3] investigated the adhesion molecules

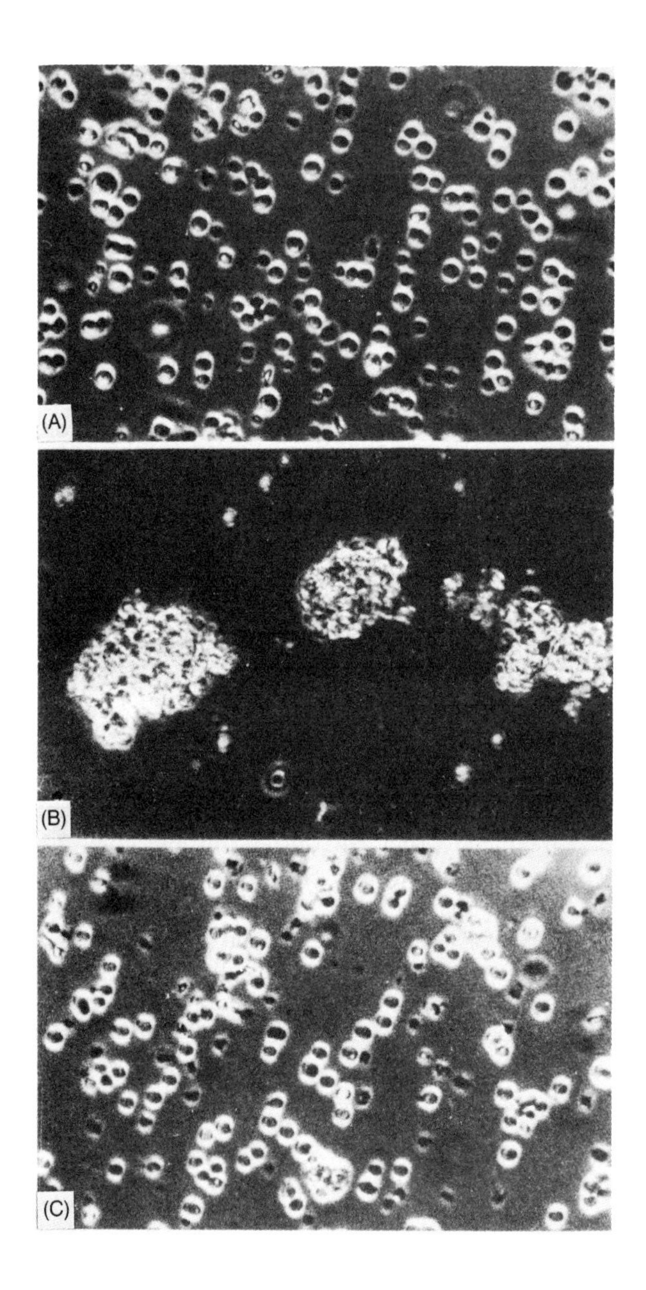

Figure 5.1 Inhibition of retinal cell aggregation by Fab fragment from anti-retinal cell antibodies. (a) Cells prior to aggregation; (b) aggregates produced after incubation for 30 min at 37°C; (c) aggregation for 30 min at 37°C with the Fab fragment. (Reprinted from [3].) © ASBMB.

involved in the selective adhesion of chick embryonic retina cells. Adhesion of trypsin-released neural retina cells was assayed by rotating cultures and determining the decrease in the number of particles. Molecules mediating the adhesion were probed by the antibody inhibition method. Rabbits were immunized with embryonic retina cells. Monovalent fragments of immunoglobulins in the antiserum (Fab fragments) inhibited the adhesion reaction (Figure 5.1), showing that determinants required for adhesion were recognized by the antibodies [3]. However, anti-retina serum contained many antibodies and identification of the molecule recognized by the inhibitory antibody required extensive work. A component of the neutralizing antigen released into the tissue culture supernatant, possibly by proteolysis, was purified by ion-exchange chromatography, gel filtration and gel electrophoresis [4]. The purified component was used to raise specific antibodies against the adhesion molecule. In order to obtain an intact adhesion molecule, neural retina was dissolved in detergent, and the adhesion molecule, N-CAM, was purified by affinity chromatography using the specific antibody. N-CAM thus obtained from the embryonic source was composed of glycoproteins of molecular weight 250–200 K, while that from the adult source was molecular weight 160–110 K [5]. The size difference between the embryonic and the adult glycoproteins is due to difference in carbohydrate (Chapter 6).

When isolated N-CAM was incorporated into the liposome, the resulting liposome could adhere to retina cells, and the adhesion was blocked by the anti-N-CAM Fab fragment [6]. This experiment demonstrated the direct role of N-CAM in cell adhesion.

Knowing the partial amino acid sequence of isolated N-CAM and utilizing specific antibodies reacting with N-CAM, the cDNA clone of N-CAM was isolated and its complete structure elucidated [7].

Three forms (ld, sd, ssd) of N-CAM are known. ld and sd are transmembrane glycoproteins, and ld has a longer cytoplasmic tail than sd. ssd is bound to the membrane by the phosphatidyl inositol anchor. The extracellular portions of the three types of N-CAM molecules are identical and contain five immunoglobulin-like domains (Figure 5.2) [7]. Comparison of the number of amino acids between the conserved cysteine residues (50–56 residues), shows that the domain in N-CAM is similar to C domain. However, the conservation of the D–X–A(G)–X–Y–X–C sequence around the second cysteine makes it more similar to the V domain. Thus, the Ig-like

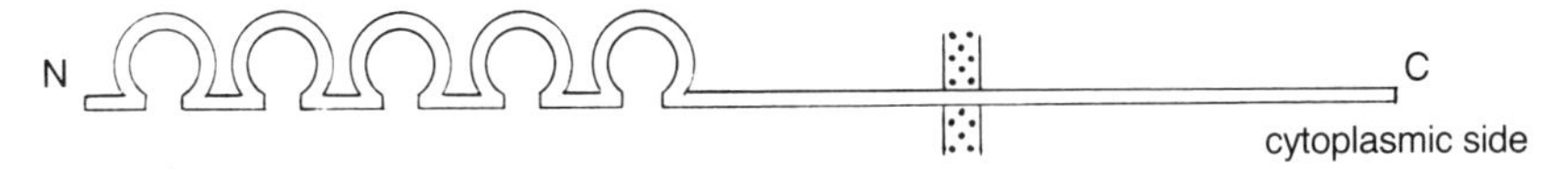

Figure 5.2 Model of N-CAM. The largest one (ld) is shown. The loops are immunoglobulin-like domains. (Modified from [7].)

domain of N-CAM might have evolved from a common ancestor before the evolutionary distinction of the V and C domains.

N-CAM accomplishes cell adhesion by homophilic (N-CAM to N-CAM) association, and the domain–domain interaction of the immunoglobulin-like domains should play an important role in the association. N-CAM also interacts with heparan sulphate-proteoglycan; through the interaction, N-CAM may also participate in cell–matrix adhesion [8].

One of the major functions of N-CAM may be to modulate other cell surface recognition interactions by drawing two adjacent cells closer and promoting interaction. Choline acetyltransferase activity of embryonic chick sympathetic neurons cultured *in vitro* increases upon cellular contact. The contact-dependent increase of the enzymatic activity is inhibited by the anti-N-CAM Fab fragment [9].

N-CAM might also participate in the formation of the neuromuscular junction. It is detected both in the neurons and muscle cells forming the junction, and the adhesion of nerve cells and muscle cells is inhibited by the anti-N-CAM Fab fragment [10].

Immunohistochemical studies have shown that N-CAM undergoes complex modes of appearance and disappearance during embryogenesis [11]. The temporally intense expression of N-CAM during embryonic induction suggests that it is involved in some way in the process.

5.1.2 Myelin-associated glycoprotein and L1

In addition to N-CAM, two proteins involved in the adhesion of nerve cells, namely myelin-associated glycoprotein [12] and L1 [13], belong to immunoglobulin superfamily.

Myelin-associated glycoprotein (MAG) was originally found as a glycoprotein (100 K) characteristically expressed in myelin [14]. MAG is expressed in oligodendrocytes prior to myelination, and appears to be involved in the process [15]. Anti-MAG autoantibody is detected in the sera of some patients with peripheral neuropathies and is believed to be responsible for demyelination [16]. MAG can bind to collagens and heparin, but not to fibronectin, laminin and N-CAM [17]. This molecule appears to accomplish cell adhesion both by a homophilic association and specific recognition.

L1 is involved in neuron–neuron adhesion, neurite fasciculation, the outgrowth of neurites and other cellular activities. Ng-CAM is probably identical to L1. L1 has 6 immunoglobulin-like domains per molecule, although N-CAM and MAG have 5 such domains [13]. In addition, L1 has regions homologous to the fibronectin repeating unit. Furthermore, two RGD sequences (the fibronection receptor-binding domain) are found in an immunoglobulin-like domain. Because of the complexity, L1 appears to be involved in multiple molecular interactions at the cell surface.

5.1.3 CD antigens and ICAM-1

Some of the cell adhesion molecules involved in lymphoid cell adhesion are also members of the immunoglobulin superfamily.

CD4 is a cell surface marker expressed in a subpopulation of T cells, and is a member of the immunoglobulin superfamily (Chapter 3). As has been described, a monoclonal antibody against CD4 inhibits the recognition of the target cell by the $CD4^+$ cell. This observation has led to a proposal that CD4 and MHC class II molecules recognize each other and serve as cell adhesion molecules. This point has been proven by a gene transfer experiment (Figure 5.3) [18]. When CV1 fibroblast cells are transfected with CD4 cDNA, the transfectants adhere to Raji B cells bearing the Class II antigen, but not to mutant cells lacking the antigen. The adhesion reaction is specifically inhibited by antibodies against CD4 or Class II antigen. Since Class II antigen is also a member of the immunoglobulin superfamily, the domain–domain interaction of the immunoglobulin-like domains may be important in the interaction. Cytotoxic T cells express a cell surface marker, CD8, which is also a member of the immunoglobulin superfamily (Chapter 3). CD8 reacts with MHC Class I antigen on target cells, and plays a role very similar to CD4 in target cell recognition.

CD2 is a cell surface marker expressed on all T cells and is a member of the immunoglobulin superfamily (Chapter 3). Monoclonal antibodies against CD2 inhibit the action of cytotoxic T cells by interfering with the adhesion of T cells to the target cells. The ligand of CD2 is LFA-3 (lymphocyte function associated antigen-3) [19]. LFA-3 (45–66 K) is found in most tissues

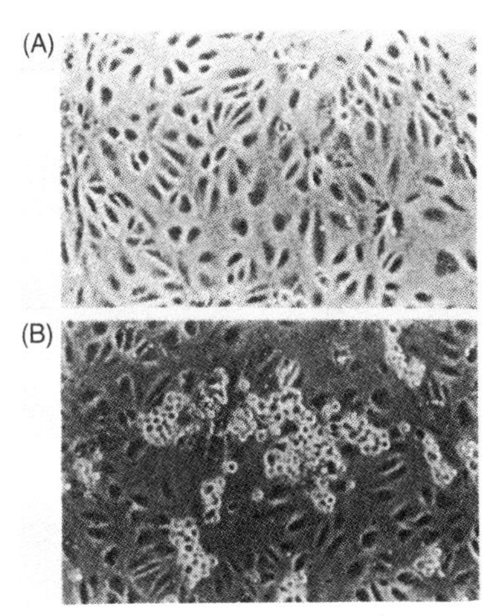

Figure 5.3 Raji B lymphocytes which expressed MHC class II antigen did not bind to monolayer of CV1 cells (A), but bound to CV1 cells, which came to express CD4 by gene transfer (B). (Reprinted from [18].) © 1987. Macmillan Magazines Ltd.

examined. The sequence of LFA-3 protein is not homologous to any other protein so far determined [20].

ICAM-1 (inter cellular adhesion molecule-1; 76–114 K) is also a member of the immunoglobulin superfamily [21, 22] and is the ligand of LFA-1 (lymphocyte function associated antigen-1) [23], which is expressed in lymphoid cells and belongs to the integrin superfamily. ICAM-1 is expressed in fibroblasts, endothelial cells and keratinocytes. The interaction of LFA-1 and ICAM-1 is important in the function of cytotoxic T cells, helper T cells, and in antibody-dependent killing by natural killer cells and granulocytes.

Thus, many cell adhesion molecules are involved in the recognition of target cells by T cells, and the majority belong to immunoglobulin superfamily. Most adhesion molecules of the immunoglobulin superfamily are involved in cell–cell adhesion, while MAG and N-CAM appear to be also involved in cell–matrix adhesion.

5.2 CADHERIN FAMILY

Cadherins are a family of proteins involved in calcium-dependent cell to cell adhesion. Takeichi has shown that intercellular adhesion in a variety of cell

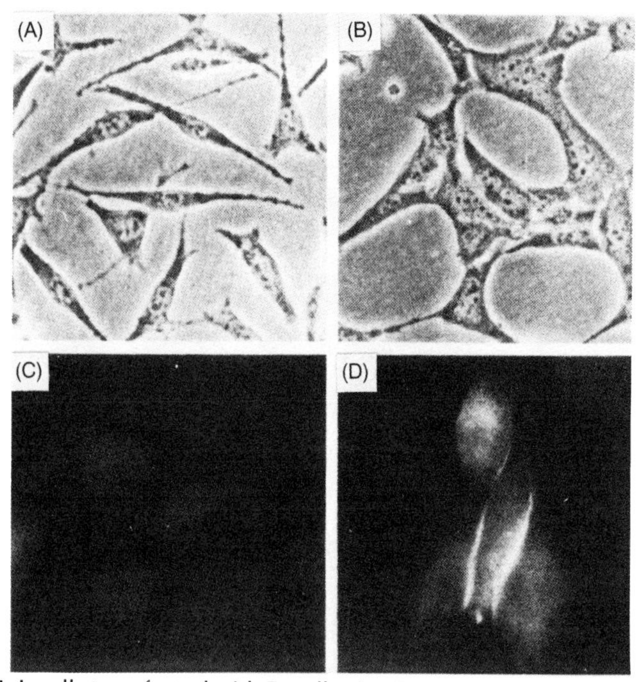

Figure 5.4 L-cells transfected with E-cadherin cDNA came to express E-cadherin and to adhere closely. (a), (c) before transfection; (b), (d) after transfection. (c), (d) staining with an anti-E-cadherin monoclonal antibody. (Reprinted from [27].) © 1987. Macmillan Magazines Ltd.

types is governed by at least two distinct sites, a Ca^{2+} dependent site and a Ca^{2+} independent site [24]. When V79 lung cells are treated with trypsin in the presence of Ca^{2+}, the Ca^{2+}-independent site is destroyed, but the Ca^{2+}-dependent site remains. Thus, the treated cells can reaggregate in the presence of Ca^{2+} but not in the absence of Ca^{2+}. N-CAM does not require Ca^{2+} for adhesion: N-CAM as well as other members of the immunoglobulin superfamily belong to the Ca^{2+}-independent adhesion molecule. The principal cell adhesion system in many cells appears to be the Ca^{2+}-dependent one, since some monoclonal antibodies, disrupting only the Ca^{2+}-dependent adhesion system, can disrupt the adhesion of these cells.

The prototype of the Ca^{2+}-dependent adhesion molecule is L-CAM. Edelman and co-workers have studied the cell adhesion molecule of chicken liver cells by the antibody inhibition method. L-CAM, the adhesion molecule thus revealed, is a transmembrane protein of molecular weight 124 K [25]. The Ca^{2+}-dependent adhesion system is important in the compaction of early mouse embryos. Jacob and co-workers used EC cells to elicit antibodies reacting with the adhesion molecule on early embryos [26]. The molecule thus disclosed is called uvomorulin or E-cadherin (E = epithelial) and is the mouse homologue of L-CAM.

The direct role of these molecules in cell adhesion has been demonstrated by a gene transfer experiment involving E-cadherin. When L fibroblasts, which do not express E-cadherin and do not form a tight intercellular connection, are transfected with E-cadherin cDNA, the resultant cells come to adhere tightly to each other (Figure 5.4) [27]. E-cadherin and L-CAM are believed to achieve cell adhesion by homophilic associations. Ca^{2+} must play an important role in the association.

Takeichi and co-workers have prepared a monoclonal antibody inhibiting the Ca^{2+}-dependent adhesion of embryonic brain cells [28] and another one inhibiting the adhesion of visceral endoderm cells [29]. These antibodies define two distinct members of the cadherin family, namely N-cadherin (N = nerve) and P-cadherin (P = placenta) (Table 5.1). All three

Table 5.1 Molecules of cadherin family

Name	Molecular weight (K)	Example of expressing tissues
E-cadherin (L-CAM, uvomorulin)	124	Epithelium
P-cadherin	118	Placenta
N-cadherin (A-CAM)	127	Nervous system

```
                                        ▼
P   MELLSGPHAFLLLLLQVCWLRSVVSEP---------------------------------   27
    *         *******        **
E   MGARCRSFSALLLLLQVSSWLCQELEPESCSPGFSSEVYTFPVPERHLERGHVLGRVRFE   60

P   ------------------------YRAGFIGEAGVTLEVEGTDLEPSQVLGKVALAGQGM   63
                                                  *
E   GCTGRPRTAFFSEDSRFKVATDGTITVKRHLKLHKLETSFLVRARDSSHRELSTKVTLKS  120

P   HHADNGDIIMLTRGTVQGGKDAMHSPPTRILRRRKREWVMPPIFVPENGKGPFPQRLNQL  123
                          *       *** ** ** ***  *** ** ** *  * **
E   MGHHHHRHHHRDPASESNPELLMFPSVYPGLRRQKRDWVIPPISCPENEKGEFPKNLVQI  180

P   KSNKDRGTKIFYSITGPGADSPPEGVFTIEKESGWLLLHMPLDREKIVKYELYGHAVSEN  183
    *** *  ** ****** *** ** *** ** * ***    ***** * ** ** **** *
E   KSNRDKETKVFYSITGQGADKPPVGVFIIERETGWLKVTQPLDREAIAKYILYSHAVSSN  240

P   GASVEEPMNISIVVTDQNDNKPKFTQDTFRGSVLEGVMPGTSVMQVTATDEDDAVNTYNG  243
    *  ** ** * * ******* * ***  * *** **  ****** * *** ** ***** 
E   GEAVEDPMEIVITVTDQNDNRPEFTQEVFEGSVAEGAVPGTSVMKVSATDADDDVNTYNA  300

P   VVAYSIHSQEPKEPHDLMFTIHKSTGTISVISSGLDREKVPEYRLTVQATDMDGEGSTTT  303
      ** * ** *  **  ***    ** ***  ******  * * * *** *  ***  **
E   AIAYTIVSQDPELPHKNMFTVNRDTGVISVLTSGLDRESYPTYTLVVQAADLQGEGLSTT  360

P   AEAVVQILDANDNAPEFEPQKYEAWVPENEVGHEVQRLTVTDLDVPNWPAWRATYHIVGG  363
    * **    * ***** * *  *   ******      * *** * ** ***      *  
E   AKAVITVKDINDNAPVFNPSTYQGQVPENEVNARIATLKVTDDDAPNTPAWKVVYTVVND  420

P   DDGDHFTITTHPETNQGVLTTKKGLDFEAQDQHTLYVEVTNEAPFAVKLPTATATVVVHV  423
     *         * ** * * * *******  *  * * * ** **   *   **** * *
E   PDQQ-FVVVTDPTTNDGILKTAKGLDFEAKQQYILHVRVENEEPFEGSLVPSTATVTVDV  479
```

```
P  K D V N E A P V F V P P S K V I E A Q E G I S I G E L V C I Y T A Q D P D K E – D Q K I S Y T I S R D P A N W L A V D P   482
     * * * * * *   *   *           *     *         *           * * *     * *       * * * *   *   *   * *   * * * *       *
E  V D V N E A P I F M P A E R R V E V P E D F G V G Q E I T S Y T A R E P D T F M D Q K I T Y R I W R D T A N W L E I N P   539

P  D S G Q I T A A G I L D R E D E Q F V K N N V Y E V M V L A T D S G N P P T T G T G T L L L T L T D I N D H G P I P E P   542
       *   *             * * * *       * * *     *           * * *   *   *     * * * * * * * *   *   *   * *     * * * * *
E  E T G A I F T R A E M D R E D A E H V K N S T Y V A L I I A T D D G S P I A T G T G T L L L V L L D V N D N A P I P E P   599

P  R Q I I I C N Q S P V P Q V L N I T D K D L S P N S S P F Q A Q L T H D S D I Y W M A E V S E – K G D T V A L S L K K F   601
   *         *       *   *         *   *   * *   * *   * * *   *   * * *           *     *                     *       *
E  R N M Q F C Q R N P Q P H I I T I L D P D L P P N T S P F T A E L T H G A S V N W T I E Y N D A A Q E S L I L Q P R K D   659

P  L K Q D T Y D L H L S L S D H G N R E Q L T M I R A T V C D C H G Q V F N D C P R – P W K G – G F I L P – I L G A V – –   656
   *         *     * *   *   *     *     *   *           * * * *   *   *   *   *                 *       *   * * *
E  L E I G E Y K I H L K L A D N Q N K D Q V T T L D V H V C D C E G T V N N – C M K A G I V A A G L Q V P A I L G I L G G   718

P  – L A L L T L L L A L L L L V R K K R K V K E P L L L P E D D T R D N V F Y Y G E E G G G E E D Q D Y D I T Q L H R G L   715
     * * * *   *   *   * * *     *         * * * * * *   *   * * * * * * *   * *   * * * * * * * * * *   *     * * * * * *
E  I L A L L I L I L L L L L F L R R R T V V K E P L L P P D D D T R D N V Y Y Y D E E G G G E E D Q D F D L S Q L H R G L   778

P  E A R P E V V L R N D V V P T F I P T P M Y R P R P A N P D E I G N F I I E N L K A A N T D P T A P P Y D S L M V F D Y   775
     * * * * *     * * * *   * *         *   * * * * * * * * * * * * * * *   * * * * * *     * * * * * * * * * *   * * * *
E  D A R P E V T – R N D V A P T L M S V P Q Y R P R P A N P D E I G N F I D E N L K A A D S D P T A P P Y D S L L V F D Y   837

P  E G S G S D A A S L S S L T T S A S D Q D Q D Y N Y L N E W G S R F K K L A D M Y G G G E D D   822
   * * * * *   * * * * * * *     *   * * * * * * *   * * * * * *   * * * * * * * * * * * * * * *
E  E G S G S E A A S L S S L N S S E S D Q D Q D Y D Y L N E W G N R F K K L A D M Y G G G E D D   884
```

Figure 5.5 Comparison of P- and E-cadherin sequences. Identical residues are marked by asterisks. The signal sequence and the transmembrane domain are overlined. The 4 conserved cysteines are enclosed in boxes. The putative N-terminus of the mature protein is marked by an arrow. (Cited from [30].)

cadherins so far detected (E, N and P) have similar molecular weights, and the protein sequences deduced from the cDNA sequences have homology of around 50% (Figure 5.5) [30–32]. These results have established that Ca^{2+}-dependent cell–cell adhesion is mediated by closely related but distinct molecules of the cadherin family. Electron microscopical studies have shown that cadherins are rich in so-called adherent junctions. The homology between cadherin molecules is most extensive in the cytoplasmic domain. Immunofluorescent studies have suggested that the cytoplasmic domains of cadherins are associated with actin filaments [33]. The homologous cytoplasmic domain may be required for the association, and this association may be needed for cell adhesion. Indeed, E-cadherin lacking its cytoplasmic tail loses cell adhesion activity [34].

Differences in cadherin subclasses appear to be the basis of selective cell adhesion, as seen in classical *in vitro* aggregation experiments, since cadherins associate with the same cadherin but not significantly with cadherins of different subclasses [35]. Takeichi and co-workers studied the fate of cadherin molecules during embryogenesis and obtained the following results [32]. Preimplantation mouse embryos express only E-cadherin. During implantation, P-cadherin starts to appear in the extraembryonic region. The embryonic ectoderm continues to express E-cadherin even after implantation. However, in the gastrulation step, the mesoderm and the definitive endoderm cease to express E-cadherin and begin to express N-cadherin. Thus, in cell populations segregating from the ectoderm, E- to N-conversion of cadherins takes place. This is also the case upon neurulation. Mesodermal cells of the somite express N-cadherin, while they lose N-cadherin when they are transformed to mesenchymal cells and leave the somite. These and other results demonstrate that the separation of a cell layer into two accompanies a class change of cadherins. The following observations, furthermore, indicate that two cell groups scheduled to connect with each other express identical cadherins. Thus, the dorsal root ganglia cells start to express N-cadherin at the time when they connect with N-cadherin-expressing neural tube. Migrating primordial germ cells arrive and stop at the genital ridge. Both the primordial germ cells and the genital ridge express N-cadherin. Although mesonephric tube and Wolffian duct express different cadherins, the two tissues express common cadherin subclasses at the site where they fuse. Furthermore, a gene transfer experiment indicates that N-cadherin is involved in guided extension of neurites [36].

As above, cadherins are one of the key molecules regulating morphogenesis. Tight association of cytoplasmic components with the cytoplasmic side of cadherins may indicate that some of the extracellular information regulating differentiation is also transmitted by cadherins.

5.3 INTEGRIN SUPERFAMILY

5.3.1 Role of fibronectin in embryogenesis and differentiation

Matrix components have been known to play significant roles in differentiation and development. In 1966, Hauschka and Konigsberg demonstrated that the treatment of a culture dish with collagen was required for clonally cultured myoblasts to fuse and produce myotubules [37]. More recent studies have shown the importance of fibronectin, which recognizes both matrix components such as collagens, and the cell surface. Thiery and co-workers examined the role of fibronectin by combining various approaches. Neural crest cells, which are formed as cell aggregates present on the top of the neural tube, migrate to a series of specific sites and form a variety of differentiated tissues such as the peripheral nervous system [38]. The pathway where the chick neural crest cells migrate is rich in fibronectin, as revealed by immunohistochemical staining. Monovalent antibodies against fibronectin prevent the migration of neural crest cells. When injected into an amphibian (*Pleurodeles waltlii*) embryo, anti-fibronectin monovalent antibodies inhibit gastrulation but not neurulation [39]. An oligopeptide, which corresponds to the cell binding sequence of fibronectin (a sequence containing RGD), also inhibits gastrulation in amphibian embryos and neural crest cell migration in chick embryos [40]. All these results firmly establish that fibronectin is involved in the migration of neural crest cells and gastrulation.

When MEC erythroleukemia cells are induced to differentiate by treatment with DMSO, cells in suspension differentiate into erythroblasts, but do not enucleate. When the cells are similarly treated in dishes coated with fibronectin, cells differentiate into enucleated cells (reticulocytes and erythrocytes) [41]. In the normal differentiation of blood cells, the whole process of differentiation occurs in the stroma of the bone marrow. Thus, fibronectin in the stroma appears to be important in erythrocyte differentiation.

5.3.2 Fibronectin receptor

Fibronectin exerts its effect by binding to a specific cell surface receptor. Studies on the molecular nature of the receptor proceeded by two different approaches. Buck and co-workers employed the antibody-inhibition approach and identified an antigen involved in the cell–substratum adhesion of chick embryonic fibroblasts. The structure of a polypeptide of the antigenic protein has been elucidated from cDNA cloning and it has been named integrin [42]. The name comes from the concept that integrin, a transmembrane polypeptide, binds to both the matrix component and the

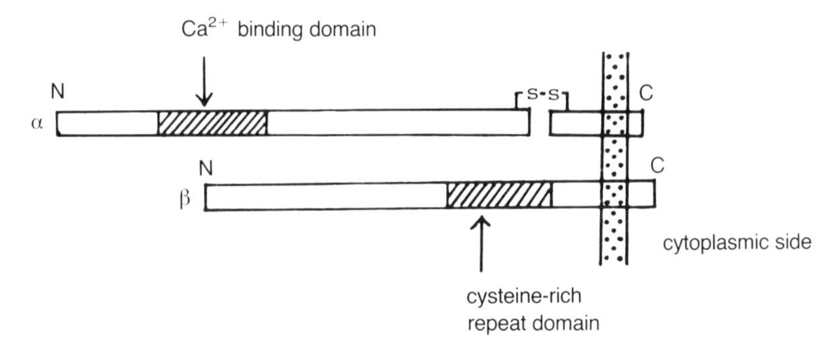

Figure 5.6 Structural model of members of integrin superfamily (based on [46]).

cytoplasmic component, and integrates the extracellular structure and the intracellular structure. Integrin apparently binds both fibronectin and laminin. Only a little is known about the linkage of these receptors to cytoskeletons. Talin has been implicated as a link protein between integrin and cytoskeletons [43].

Ruoslahti and co-workers utilized their finding that the Arg–Gly–Asp (RGD) sequence in fibronectin was the receptor binding site for fibronectin. They isolated the receptor from a detergent extract of osteosarcoma cells by affinity chromatography on fibronectin–agarose. RGD peptide was used to specifically elute the receptor from the affinity column. Fibronectin receptor thus isolated reacted only with fibronectin and not with laminin or vitronectin. The receptor consists of two non-covalently associated subunits, α (160 K) and β (120 K) [44]. The structures of both subunits have been established by cDNA cloning (Figure 5.6) [45]. The α-subunit contains a motif called EF hand, which is found in many calcium binding proteins (Figure 5.7). The subunit is post-transcriptionally cleaved into a large (~ 140 K) and a small polypeptide (20 K). The β-subunit corresponds to integrin, and has a domain rich in cysteine (Figure 5.6). Upon SDS gel electrophoresis, both α- and β-subunits show a characteristic behaviour.

E	F	S	G	D	D	T	E	D	F	V	A	fibronectin receptor site 1
D	V	N	G	D	G	L	D	D	L	L	V	site 2
D	R	T	P	D	G	R	P	Q	E	V	G	site 3
D	L	D	Q	D	G	Y	N	D	V	A	I	site 4
D	L	D	G	N	G	Y	P	D	L	I	V	site 5
D	Q	N	R	D	G	I	I	D	–	–	D	myosin light chain
D	–	DN	–	D	G	–	I	D	K	D	D	troponin C consensus
D	–	D	G	DN	G	–	I	–	–	–	E	calmodulin consensus

Figure 5.7 Structure of Ca^{2+}-binding domain of the α-subunit of fibronectin receptor. Boxes show amino acids commonly found in other Ca^{2+}-binding proteins. (Based on [45].)

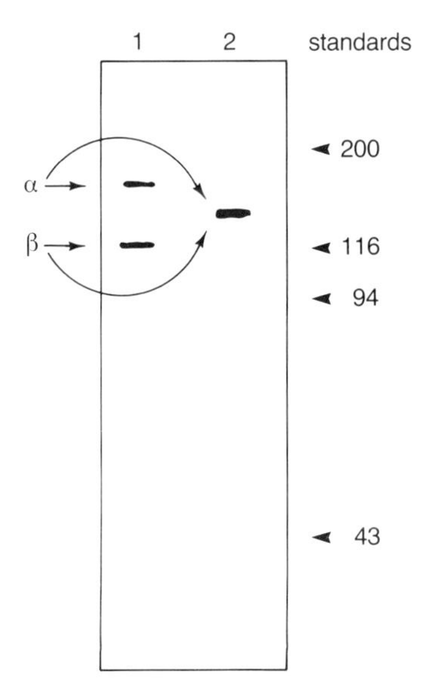

Figure 5.8 Behaviour of fibronection receptor upon SDS gel electrophoresis. 1: Under non-reducing conditions; and 2: under reducing conditions. (Based on [46].)

Under reducing conditions, α-subunit migrates faster as compared to that under non-reducing conditions, because of the separation of the large and the small peptides (Figure 5.8). On the other hand, the β-subunit migrates more slowly under reducing conditions due to a change in shape caused by the intrachain –S–S– cleavage.

Fibronectin receptor is the prototype of a large superfamily of proteins involved in cell adhesion, principally in cell–substratum adhesion [46]. The protein superfamily is called the integrin superfamily. Vitronectin receptor recognizes vitronectin, a matrix protein, and the RGD sequence in vitronectin is also the receptor binding site. The receptor is also composed of an α-subunit and a β-subunit, both of which are homologous to the subunits of fibronectin receptor [47]. Fibrinogen receptor in platelets is required for the initial step of blood coagulation. The receptor is platelet glycoprotein IIIa [46, 48]. Its β-subunit appears to be identical to vitronectin receptor.

Since the members of the integrin family directly connect the matrix components with cytoskeleton, some information critical to differentiation may be transmitted through the receptor. Indeed, this appears to be the case for myogenesis. When chick embryonic myoblasts cultured on collagen-coated dishes are treated with a monoclonal antibody reacting with the β-subunit of fibronectin receptor, myoblasts continue to replicate but do not fuse or produce muscle specific myosin (Table 5.2) [49]. The requirement

Table 5.2 A monoclonal antibody to the β-subunit of integrin (Ab) inhibits myogenesis of chicken embryonic myoblasts

	Days in culture		Nuclei in myotubes (%)
	+Ab	−Ab	
Experiment 1			
Control	0	6	72
Ab-treated	6	0	3
Ab-reversed	3	3	66
Ab-reversed	6	4	65
Experiment 2			
Ab-treated	6	0	4
Ab-reversed	3	3	75

Based on [49].

for collagen in myogenesis is apparently to establish an environment allowing fibronectin to interact with the receptor.

5.3.3. LFA-1 and related molecules

LFA-1, Mac-1 and Gp 150/95 are leukocyte adhesion proteins of the integrin superfamily. These proteins were principally studied by Springer and co-workers [50]. LFA-1 is defined by monoclonal antibodies inhibiting the T-cell mediated killing of target cells. LFA-1 on T cells is involved in the first step of the killing, namely adhesion of the T cells to the target cells. The ligand of LFA-1 is ICAM-1, a member of the immunoglobulin superfamily. Mac-1 is a differentiation antigen expressed on macrophages and granulocytes but not on lymphocytes. Mac-1 is the complement receptor type 3 (CR3) and is involved in complement-dependent adhesion and phagocytosis of particles to which the cleaved third component of complement is attached. Gp 150/95 is expressed in monocytes and granulocytes. The β-subunit (95 K) of all these molecules appears to be identical, while the α-subunits (180–150 K) are different. Indeed the sequence of the β-subunit deduced from the cDNA sequence shows 45% homology with integrin (Figure 5.9) [51]. The deficiency of the β-subunit in human results in an inheritable disease characterized by recurrent bacterial infections [52].

5.3.4 PS antigens in *Drosophila*

A group of proteins similar to fibronectin receptor is involved in cell differentiation in *Drosophila*. They are position-specific (PS) antigens. In the organism, much of the adult epidermis derives from the larval imaginal disks. Monoclonal antibodies have disclosed two antigens, one preferentially expressed in cells of the dorsal compartments of the mature wing imaginal disk and the other one preferentially expressed in the ventral compartments [53]. The former antigen is called PS1, and the latter is PS2 antigen. PS1 antigen is on a polypeptide of 120 K and PS2 on a 125 K polypeptide. Both PS antigens are complexed with a common subunit of 110 K. A monoclonal antibody reacting with the common subunit discloses that:

1. In addition to PS1 and PS2 antigens, at least one other PS antigen of 92 K is present.
2. PS antigens become intensely expressed after the onset of gastrulation.
3. Even adult tissues express PS antigens [54].

The common component of the PS antigen (110 K) is reminiscent of the β-subunit of integrin superfamily, while the variable subunit appears to correspond to the α-subunit. Indeed, the N-terminal sequence of one of the variable units is homologous to the α-subunit [55]. The position specific antigen appears to be involved in a variety of processes of differentiation and development, particularly cell adhesion, cell migration and even in differentiation. Since the molecular genetic analysis of embryogenesis is much advanced in *Drosophila*, key information on the function of the integrin superfamily in development should come from the studies on the PS antigens.

5.3.5 Very late antigens

Another developmentally regulated antigen of the integrin superfamily is VLA (very late antigens) found on long-term activated T cells. The antigens are expressed on T cells after *in vitro* stimulation for 2–4 weeks. Thus far, five antigens (VLA1–VLA5) are known; they share a common β-subunit (130 K) and have unique α-subunits [56, 57]. Antibody blocking studies indicate that the antigen is involved in cell adhesion to fibronectin and laminin. Both subunits of VLA are immunologically related to the subunits of the fibronectin receptor [57].

5.4 COMMENTS

In addition to the 3 groups of proteins mentioned in this chapter, proteins belonging to other families are probably involved in cell adhesion, since there are still many cell adhesion factors which are ill defined in molecular terms. Furthermore, glycosaminoglycan chains in cell surface proteoglycans are certainly involved in cell adhesion, since the carbohydrate chains are recognized by many proteins, such as N-CAM, fibronectin [58] and laminin,

```
β-subunit  MLGLRPPLLALVG-LLSLGCVLS-----QECTKFKVSSCRECIESGPGCTWCQKLNFTGP   54
Band III   MARTNLTLLTWAGILCCLIWSGSAQQGGSDCIKANAKSCGECIQAGPNCGWCKKTDFLQE   60

           GDPDSIRCDTRPQLLMRGCAADDIMDPTSLAETQEDHNGGQK--------------QLSP  108
           GEPTSARCDDLAALKSKGCPEQDIENPRGSKRVLEDREVTNRKIGAAEKLKPEAITQIQP  120

           QKVTLYLRPGQAAAFNVTFRRAKGYPIDLYYLMDLSYSMLDDLRNVKKLGGDLLRALNEI  160
           QKLVLQLRVGEPQTFSLKFKRAEDYPIDLYYLMDLSYSMKDDLENVKSLGTALMREMEKI  180

           TESGRIGFGSFVDKTVLPFVNTHPDKLRNPCPNKEKECQPPFAFRHVLKLTNNSNQFQTE  220
           TSDFRIGFGSFVEKTVMPYISTTPAKLRNPCTG-DQNCTSPFSYKNVLSLTSEGNKFNEL  239

           VGKQLISGNLDAPEGGLDAMMQVAACPEEIGWRMVTRLLVFATDDGFHFAGDGKLGAILT  280
           VGKQHISGNLDSPEGGFDAIMQVAVCGDQIGWRMVTRLLVFSTDAGFHFAGDGKLGGIVL  299

           PNDGRCHLEDNLYKRSNEFDYPSVGQLAHKLAENNIQPIFAVTSRMVKTYEKLTEIIPKS  340
           PNDGKCHLENNMYTMSHYYDYPSIAHLVQKLSENNIQTIFAVTEEFQAVYKELKNLIPKS  359

           AVGELSEDSSNVVHLIKNAYNKLSSRVFLDHNALPDTLKVTYDSFCSNGVTHRNQPRGDC  400
           AVGTLSSNSSNVIQLIIDAYNSLSSEVILENSKLPKEVTISYKSYCKNGVNDTQEDGRKC  419

           DGVQINVPITFQVKVTATECI---QEQSFVIRALGFTDIVTVQVLPQCECRCRDQSRDRS  457
           SNISIGDEVRFEINVTANECPKKGQNETIKIKPLGFTEEVEIHLQFICDCLCQSEGEPNS  479
```

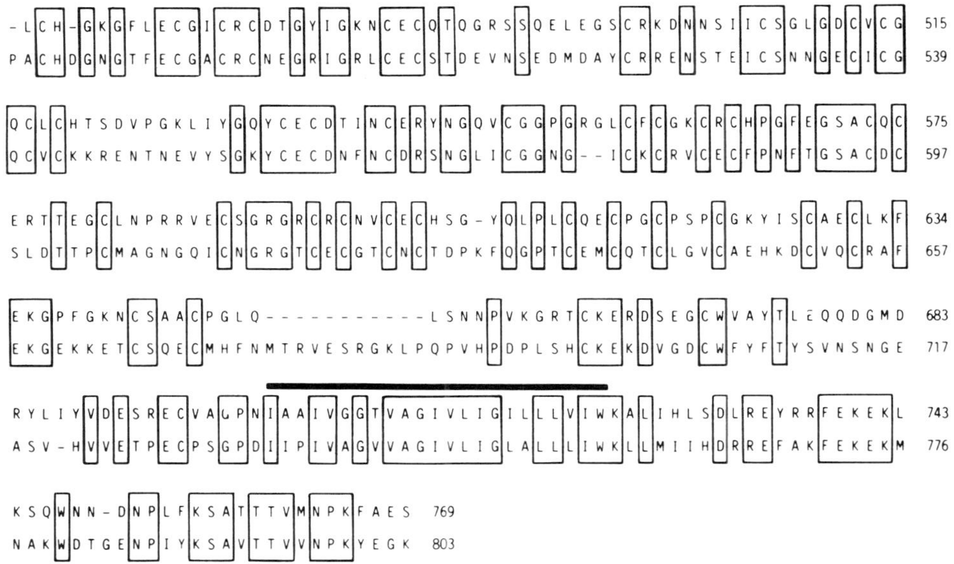

```
- L C H - G K G F L E C G I C R C D T G Y I G K N C E C Q T Q G R S S Q E L E G S C R K D N N S I I C S G L G D C V C G   515
P A C H D G N G T F E C G A C R C N E G R I G R L C E C S T D E V N S E D M D A Y C R R E N S T E I C S N N G E C I C G   539

Q C L C H T S D V P G K L I Y G Q Y C E C D T I N C E R Y N G Q V C G G P G R G L C F C G K C R C H P G F E G S A C Q C   575
Q C V C K K R E N T N E V Y S G K Y C E C D N F N C D R S N G L I C G G N G - - I C K C R V C E C F P N F T G S A C D C   597

E R T T E G C L N P R R V E C S G R G R C R C N V C E C H S G - Y Q L P L C Q E C P G C P S P C G K Y I S C A E C L K F   634
S L D T T P C M A G N G Q I C N G R G T C E C G T C N C T D P K F Q G P T C E M C Q T C L G V C A E H K D C V Q C R A F   657

E K G P F G K N C S A A C P G L Q - - - - - - - - - - - L S N N P V K G R T C K E R D S E G C W V A Y T L E Q Q D G M D   683
E K G E K K E T C S Q E C M H F N M T R V E S R G K L P Q P V H P D P L S H C K E K D V G D C W F Y F T Y S V N S N G E   717

R Y L I Y V D E S R E C V A G P N I A A I V G G T V A G I V L I G I L L L V I W K A L I H L S D L R E Y R R F E K E K L   743
A S V - H V V E T P E C P S G P D I I P I V A G V V A G I V L I G L A L L L I W K L L M I I H D R R E F A K F E K E K M   776

K S Q W N N - D N P L F K S A T T T V M N P K F A E S   769
N A K W D T G E N P I Y K S A V T T V V N P K Y E G K   803
```

Figure 5.9 The β-subunit common to leukocyte adhesion proteins (LFA-1, Mac-1 and Gp150/95) belongs to the integrin superfamily. β subunit: subunit of leukocyte adhesion proteins; Band III: β-subunit of integrin. Conserved cysteine residues are circled. The signal sequence and the transmembrane domain are overlined. (Cited from [51].)

in plasma membranes and extracellular matrix. Cell surface carbohydrates other than proteoglycan are also involved in cell adhesion as shown below and more fully in the next chapter.

An area where cell adhesion molecules could play an important role is embryonic induction. Among the types of secondary embryonic induction the most important is epithelio–mesenchymal interaction, e.g. the interaction of the epidermis and the dermis in the developing chick. In this case, dermis (mesoderm) determines the fate of epidermis (ectoderm). Thus, when wing ectoderm is grafted onto foot dermis, scales and claw develop from the epidermis instead of wing feathers. The molecular mechanism of the interaction is poorly understood, but it is certain that cell surface molecules are involved in the interaction [59].

Involvement of many cell-surface molecules in specific cell adhesion has been revealed recently in a phenomenon called 'lymphocyte homing' [60]. Circulating lymphocytes interact with high endothelial venules of lymphoid organs, and the interaction exhibits remarkable organ specificity, resulting in the homing of a certain lymphocyte subpopulation to a specific lymphoid organ. This specific interaction is mediated by complementary cell-surface molecules on lymphocytes and high endothelial venules. Homing of mouse lymphoid cells to the intestinal Peyer's patches is inhibited by an antibody recognizing a lymphocyte surface antigen, LPAM-1; and the degree of LPAM-1 expression correlates with the binding capability to Peyer's patches. Thus, LPAM-1 is a homing receptor on lymphocytes destined to settle in Peyer's patches. Immunochemical studies have shown that LPAM-1 is very similar or identical to VLA-4α, an integrin α chain [61].

MEL-14 antigen (90 K) is a homing receptor expressed on lymphocytes capable of binding to peripheral lymph nodes. The structure of MEL-14 antigen has been clarified by molecular cloning; its protein portion (42 K) consists of a carbohydrate-binding domain, an EGF-like domain, a complement binding domain and a transmembrane domain [62]. The mature protein also contains carbohydrates and branches of a peptide called ubiquitin. The importance of the carbohydrate binding domain in the function of the homing receptor has been indicated by several experiments. Notably, mannose 6-phosphate inhibits lymphocyte binding to high endothelial venules of peripheral lymph nodes. Thus, MEL-14 antigen demonstrates clearly the physiological importance of protein–carbohydrate interactions in cellular recognition.

The third class of homing receptor, Hermes antigen (CD44) (90 K), is expressed on human lymphocytes and is involved in homing to various lymphoid organs. The antigen is homologous to proteoglycan core proteins and link proteins [63, 64].

REFERENCES

1. Townes, P.L. and Holtfreter, J. (1955) *J. Exp. Zool.*, **128**, 53–120
2. Moscona, A.A. (1952) *Exp. Cell Res.*, **3**, 535–9
3. Brackenbury, R., Thiery, J.-P., Rutishauser, U. and Edelman, G.M. (1977) *J. Biol. Chem.*, **252**, 6835–40
4. Thiery, J.-P., Brackenbury, R., Rutishauser, U. and Edelman, G.M. (1977) *J. Biol. Chem.*, **252**, 6841–5
5. Hoffman, S., Sorkin, B.C., White, P.C., Brackenbury, R., Mailhammer, R., Rutishauser, U., Cunningham, B.A. and Edelman, G.M. (1982), *J. Biol. Chem.*, **257**, 7720–9
6. Rutishauser, U., Hoffman, S. and Edelman, G.M. (1982) *Proc. Natl. Acad. Sci.*, **79**, 685–9
7. Cunningham, B.A., Hemperly, J.J., Murray, B.A., Prediger, E.A., Brackenbury, R. and Edelman, G.M. (1987) *Science*, **236**, 799–806
8. Cole, G.J., Lowey, A. and Glaser, L. (1986) *Nature*, **320**, 445–7
9. Acheson, A. and Rutishauser, U. (1988) *J. Cell Biol.*, **106**, 479–86
10. Grumet, M., Rutishauser, U. and Edelman, G.M. (1982) *Nature*, **295**, 693–5
11. Edelman, G.M., Gallin, W.J., Delouvée, A., Cunningham, B.A. and Thiery, J.-P. (1983) *Proc. Natl. Acad. Sci.*, **80**, 4384–8
12. Arquint, M., Roder, J., Chia, L.-S., Down, J., Wilkinson, D. *et al.* (1987) *Proc. Natl. Acad. Sci.*, **84**, 600–04
13. Moos, M., Tacke, R., Scherer, H., Teplow, D., Früh, K. and Schachner, M. (1988) *Nature*, **334**, 701–3
14. Quarles, R.H., Everly, J.L. and Brady, R.O. (1972) *Biochem. Biophys. Res. Commun.*, **47**, 491–7
15. Sternberger, N.H., Quarles, R.H., Itoyama, Y. and Webster, H.D. (1979) *Proc. Natl. Acad. Sci.*, **76**, 1510–14
16. Braun, P.E., Frail, D.E. and Latov, N. (1982) *J. Neurochem.*, **39**, 1261–5
17. Fahring, T., Landa, C., Pesheva, P., Kühn, K. and Schachner, M. (1987) *EMBO J.*, **6**, 2875–83
18. Doyle, C. and Stominger, J.L. (1987) *Nature*, **330**, 256–9
19. Selvaraj, P., Plunkett, M.L., Dustin, M., Sanders, M.E., Shaw, S. and Springer, T.A. (1987) *Nature*, **326**, 400–03
20. Wallner, B.P., Frey, A.Z., Tizard, R., Mattaliano, R.J., Hession, C. *et al.* (1987) *J. Exp. Med.*, **166**, 923–32
21. Staunton, D.E., Marlin, S.D., Stratowa, C., Dustin, M.L. and Springer, T. (1988) *Cell*, **52**, 925–33
22. Simmons, D., Makgoba, M.W. and Seed, B. (1988) *Nature*, **331**, 624–7
23. Makgoba, M.W., Sanders, M.E., Luce, G.E.G., Dustin, M.L., Springer, T.A. *et al.* (1988) *Nature*, **331**, 86–8
24. Takeichi, M. (1977) *J. Cell. Biol.*, **75**, 464–74
25. Gallin, W.J., Sorkin, B.C., Edelman, G.M. and Cunningham, B.A. (1987) *Proc. Natl. Acad. Sci.*, **84**, 2808–12
26. Hyafil, F., Morello, D., Babinet, C. and Jacob, F. (1980) *Cell*, **21**, 927–34
27. Nagafuchi, A., Shirayoshi, Y., Okazaki, K., Yasuda, K. and Takeichi, M. (1987) *Nature*, **329**, 341–3
28. Hatta, K., Okada, T.S. and Takeichi, M. (1985) *Proc. Natl. Acad. Sci.*, **82**, 2789–93
29. Nose, A. and Takeichi, M. (1986) *J. Cell. Biol.*, **103**, 2649–58
30. Nose, A., Nagafuchi, A. and Takeichi, M. (1987) *EMBO J.*, **6**, 3655–61

31. Hatta, K., Nose, A., Nagafuchi, A. and Takeichi, M. (1988) *J. Cell Biol.*, **106**, 873–82
32. Takeichi, M. (1987) *Trends Genet.*, **3**, 213–17
33. Hirano, S., Nose, A., Hatta, K., Kawakami, A. and Takeichi, M. (1987) *J. Cell Biol.*, **105**, 2501–10
34. Nagafuchi, A. and Takeichi, M. (1988) *EMBO J.*, **7**, 3679–84
35. Nose, A., Nagafuchi, A. and Takeichi, M. (1988) *Cell*, **54**, 993–1001
36. Matsunaga, M., Hatta, K., Nagafuchi, A. and Takeichi, M. (1988) *Nature*, **334**, 62–4
37. Hauschka, S.D. and Konigsberg, I.R. (1966) *Proc. Natl. Acad. Sci.*, **55**, 119–26
38. Le Douarin, N.M. (1984) *Cell*, **38**, 353–60
39. Boucaut, J.C., Darribére, T., Boulekbache, H. and Thiery, J.P. (1984) *Nature*, **307**, 364–67
40. Boucaut, J.-C., Darribére, T., Poole, T.J., Aoyama, H., Yamada, K.M. and Thiery, J.P. (1984) *J. Cell Biol.*, **99**, 1822–30
41. Patel, V.P. and Lodish, H.F. (1987) *J. Cell Biol.*, **105**, 3105–18
42. Tamkun, J.W., DeSimone, D.W., Fonda, D., Patel, R.S., Buck, C. *et al.* (1986) *Cell*, **46**, 271–82
43. Horwitz, A., Duggan, K., Buck, C., Beckerle, M.C. and Burridge, K. (1986) *Nature*, **320**, 531–3
44. Pytela, R., Pierschbacher, M.D. and Ruoslahti, E. (1985) *Cell*, **40**, 191–8
45. Argraves, W.S., Suzuki, S., Arai, H., Thompson, K., Pierschbacher, M.D. and Ruoslahti, E. (1987) *J. Cell Biol.*, **105**, 1183–90
46. Ruoslahti, E. and Pierschbacher, M.D. (1987) *Science*, **238**, 491–7
47. Suzuki, S., Argraves, W.S., Arai, H., Langaino, L.R., Michael, D., Pierschbacher, M.D. and Ruoslahti, E. (1987) *J. Biol. Chem.*, **262**, 14080–5
48. Fitzgerald, L.A., Steiner, B., Rall, S.C., Lo, S. and Phillips, D.R. (1987) *J. Biol. Chem.*, **262**, 3936–9
49. Menko, A.S. and Boettiger, D. (1987) *Cell*, **51**, 51–7
50. Sanchez-Madrid, F., Nagy, J.A., Robbins, E., Simon, P. and Springer, T.A. (1983) *J. Exp. Med.*, **158**, 1785–1803
51. Kishimoto, T.K., O'Connor, K., Lee, A., Roberts, T.M. and Springer, T.A. (1987) *Cell*, **48**, 681–90
52. Marlin, S.D., Morton, C.C., Anderson, D.C. and Springer, T.A. (1986) *J. Exp. Med.*, **164**, 855–67
53. Wilcox, M., Brower, D.L. and Smith, R.J. (1981) *Cell*, **25**, 159–64
54. Wilcox, M. and Leptin, M. (1985) *Nature*, **316**, 351–4
55. Leptin, M., Aebersold, R. and Wilcox, M. (1987) *EMBO J.*, **6**, 1037–43
56. Hemler, M.E., Jacobson, J.G. and Strominger, J.L. (1985) *J. Biol. Chem.*, **260**, 15246–52
57. Takada, Y., Huang, C. and Hemler, M.E. (1987) *Nature*, **326**, 607–09
58. Saunders, S. and Bernfield, M. (1988) *J. Cell Biol.*, **106**, 423–30
59. Gallin, W.J., Chuong, C.-M., Finkel, L.H. and Edelman, G.M. (1986) *Proc. Natl. Acad. Sci.*, **83**, 8235–9
60. Stoolman, L.M. (1989) *Cell*, **56**, 907–10
61. Holzmann, B., McIntyre, B.W. and Weissman, I.L. (1989) *Cell*, **56**, 37–46
62. Lasky, L.A., Singer, M.S., Yednock, T.A., Dowbenko, D., Fennie, C. *et al.* (1989) *Cell*, **56**, 1045–55
63. Stamenkovic, I., Amiot, M., Pesando, J.M. and Seed, B. (1989) *Cell*, **56**, 1057–62
64. Goldstein, L.A., Zhou, D.F.H., Picker, L.J., Minty, C.N., Bargatze, R.F. *et al.* (1989) *Cell*, **56**, 1063–72

6 Cell surface carbohydrates

6.1 BIOCHEMISTRY OF CELL SURFACE CARBOHYDRATES

6.1.1 Structure

Carbohydrates which are covalently bound to proteins or to lipids are abundant on the external surface of plasma membranes (Figure 6.1). Carbohydrates are not found on the cytoplasmic side of plasma membranes, and are therefore a major cause of the asymmetry of plasma membranes. Molecules where protein is combined with carbohydrate are called glycoproteins [1–3] and where lipids are combined with carbohydrate the molecules are known as glycolipids [4]. Most transmembrane proteins are glycoproteins. The component monosaccharide species of the oligosaccharides found in the surface of mammalian cells are few in number,

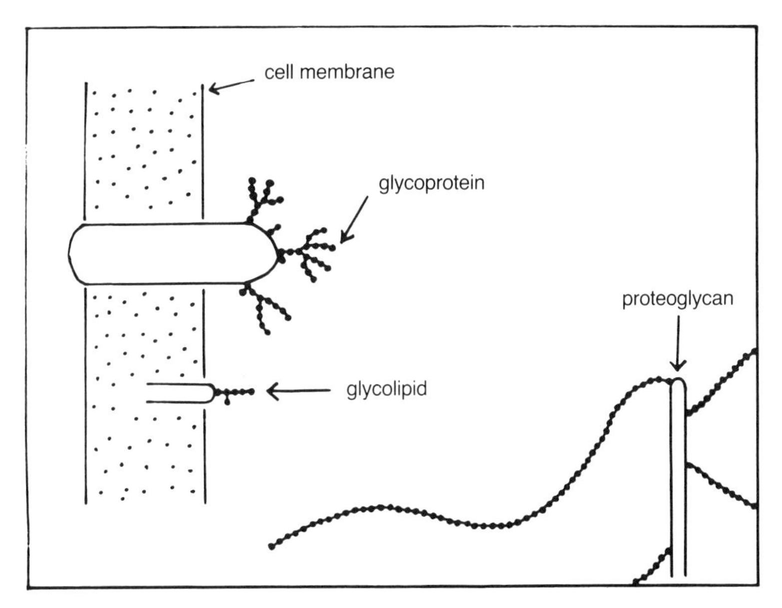

Figure 6.1 Schematic drawings of cell surface carbohydrates; black circles represent sugar residues.

Table 6.1 Monosaccharide components typically found in cell surface oligosaccharides of vertebrate cells

Name (abbreviations)	Structure
glucose (Glc)	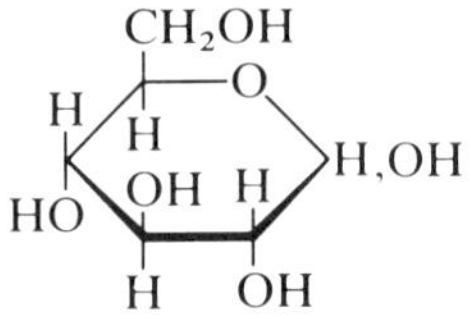 CH_2OH O H H H,OH OH H HO H OH
galactose (Gal)	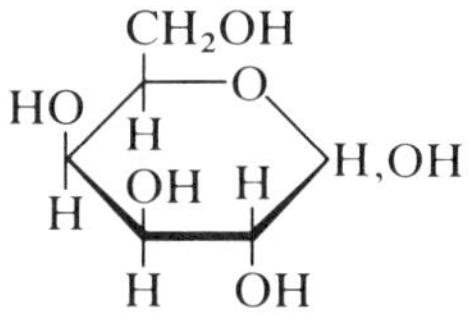 CH_2OH O HO H H,OH OH H H H OH
mannose (Man)	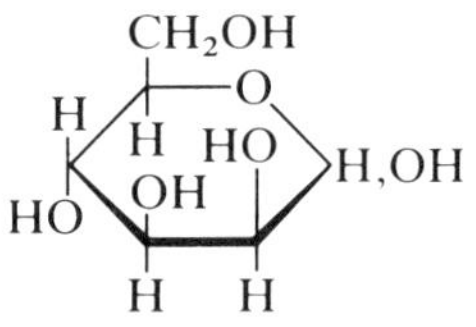 CH_2OH O H H HO H,OH OH HO H H
L-fucose (Fuc)	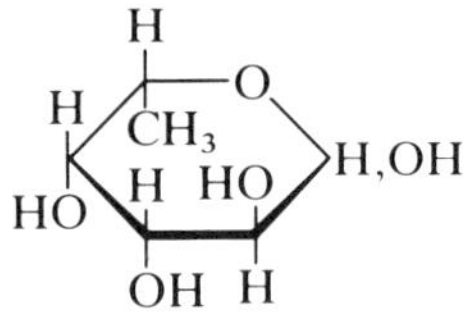 H O H CH_3 H,OH H HO HO OH H
N-acetylglucosamine (GlcNAc)	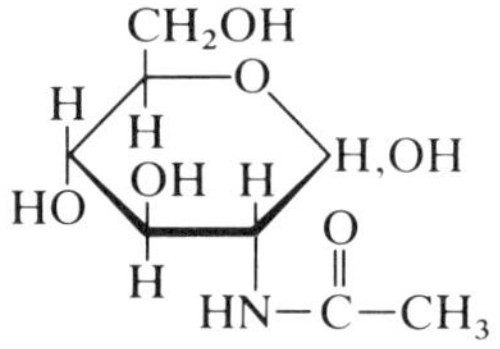 CH_2OH O H H H,OH OH H HO O H HN−C−CH_3
N-acetylgalactosamine (GalNAc)	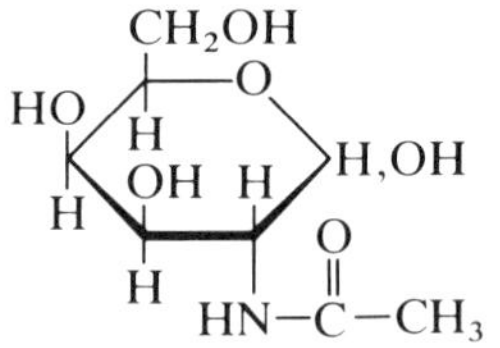 CH_2OH O HO H H,OH OH H H O H HN−C−CH_3

N-acetylneuraminic acid (NeuNAc)
(a sialic acid)

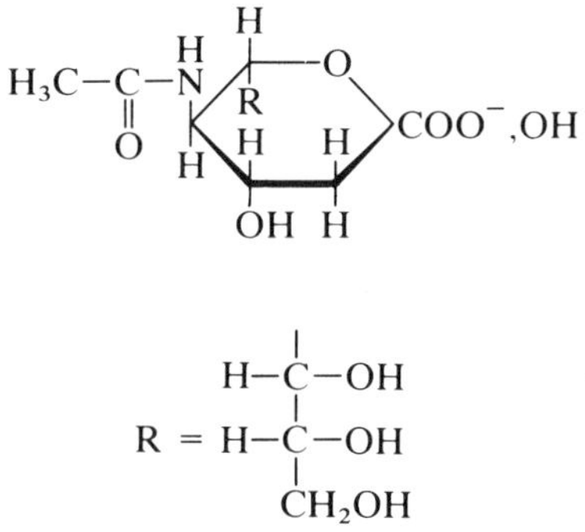

glucuronic acid (GlcA)

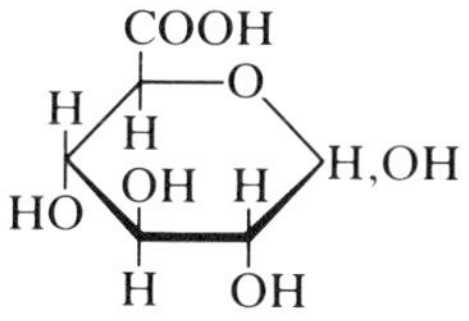

---

and the most abundant are listed in Table 6.1. However, the structures formed by these molecules can be enormously diverse. When two sugar residues are linked, numerous structures are possible; anomeric configuration (α, β) and positions of the hydroxyl groups to which neighbouring sugar residues attach are the sources of the structural diversity. Branching of the carbohydrate chain further increases the diversity.

In glycoproteins, carbohydrates are mostly linked to asparagine, serine or threonine in protein moieties. Glycoprotein-bound carbohydrates on the cell surface can be classified largely into 3 groups based on the protein–carbohydrate linkage (Figure 6.2). The first type is asparagine-linked: *N*-glycosidic linkage between *N*-acetylglucosamine and the amide group of asparagine forms the protein–carbohydrate linkage. Asparagine-linked carbohydrates are generally classified into high mannose- (or oligomannose-) type and complex- (or lactosamine-) type (Figure 6.3). The former is composed of only mannose and *N*-acetylglucosamine. In addition, the latter contains galactose, and frequently sialic acid and fucose. A common core structure composed of 3 mannosyl residues and 2 *N*-acetylglucosamine residues is present in both the high mannose-type and the complex-type. In the case of complex-type carbohydrates, side chains (antenna) of sialyl-galactosyl-*N*-acetylglucosamine sequences are usually attached to the core structure. An asparagine residue must be in the

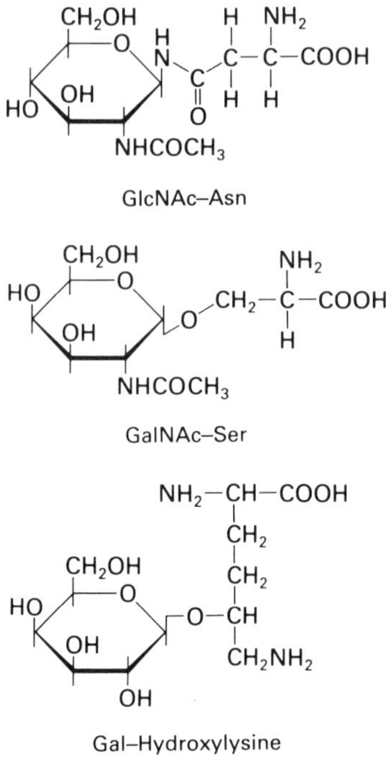

GlcNAc–Asn

GalNAc–Ser

Gal–Hydroxylysine

Figure 6.2 Examples of protein–carbohydrate linkages.

sequence of Asn–X–Thr(Ser) in order to be glycosylated. Asparagine-linked carbohydrates can be released from the peptide moiety by hydrazinolysis, and the protein–carbohydrate linkage may also be cleaved emzymatically by glycopeptidase (or *N*-glycanase). Extensive digestion with nonspecific proteases such as pronase is also a convenient method of removing most of the peptide moiety; the resulting material, called glycopeptides, usually has an intact oligosaccharide and only a few amino acids.

The second type of protein–carbohydrate linkage is *O*-glycosidic linkage between *N*-acetylgalactosamine and the hydroxyl group of serine or threonine. The carbohydrates of the linkage lack mannose, and are occasionally called 'mucin-type', since they are abundant in mucins. A variety of structures are also known for this class of carbohydrates (Figure 6.4). Treatment with mild alkali in the presence of $NaBH_4$ is the best known method of recovering the carbohydrate moiety. Since *O*-glycosidically-linked oligosaccharides are often clustered along a polypeptide chain, the glycosylated portion is sometimes resistant to proteases.

A high-mannose oligosaccharide

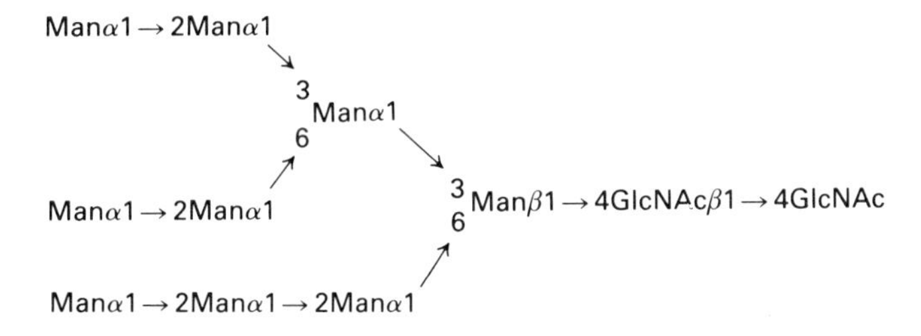

Oligosaccharides from hamster fibronectin

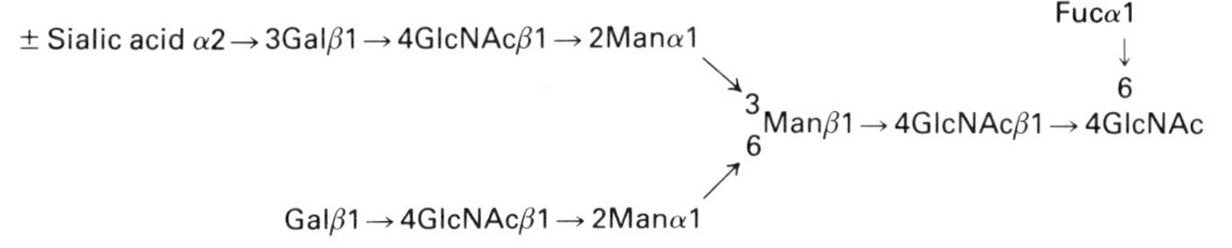

An oligosaccharide from glycophorin

(major sialoglycoprotein of erythrocytes)

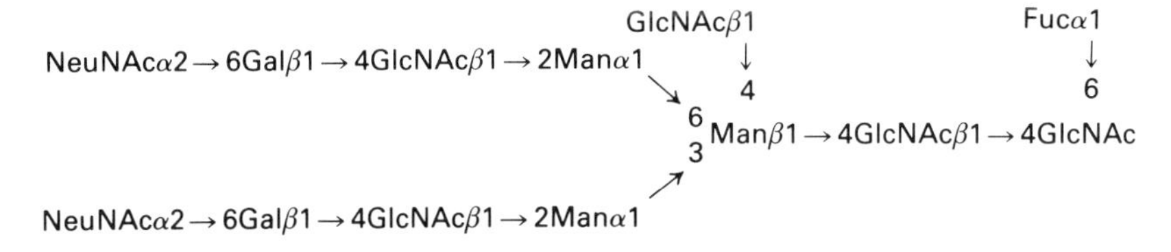

An oligosaccharide which increase in polyomavirus transformed BHK cells

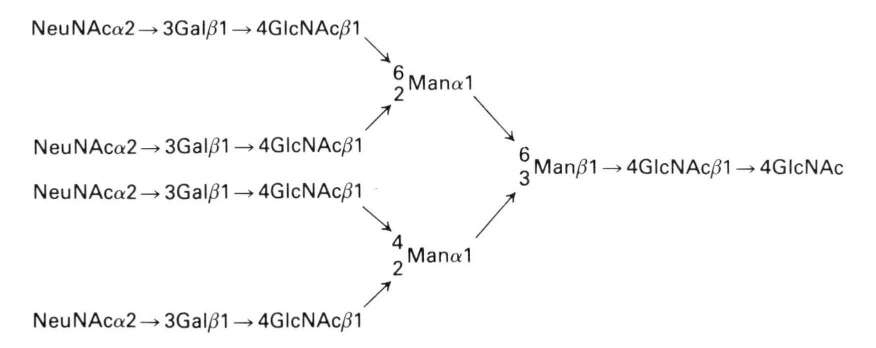

Figure 6.3 Examples of asparagine-linked oligosaccharides found in cell surface. Oligosaccharide structures found in fibronectin [82], glycophorin [83] and the virus transformed BHK cells [22] are based on respective references.

Galβ1 → 3GalNAc1
↓
6
Galβ1 → 3GalNAc

NeuNAcα2
↓
6
NeuNAcα2 → 3Galβ1 → 3GalNAc

NeuNAcα2 → 3Galβ1 → 4GlcNAcβ1
↘
6
NeuNAcα2 → 3Galβ1 → 3GalNAc

Figure 6.4 Examples of *O*-linked oligosaccharides found in cell surface.

The last type is *O*-glycosidic linkage between galactose and the hydroxyl group of hydroxylysine. This linkage connects Glcα1 → 2Gal disaccharide to collagens and related molecules. In addition, *O*-glycosidic linkages between *N*-acetylglucosamine and serine, and that between mannose and serine are known; the former is abundant in nuclear proteins [5]. Furthermore there is *O*-glycosidic linkage between xylose and the hydroxyl group of serine: this type of linkage connects many glycosaminoglycans to core proteins.

In glycolipids, the majority of the lipid carbohydrate linkage is *O*-glycosidic linkage between glucose and the hydroxyl group of sphingosine (Figure 6.5). The lipid portion of these glycolipids is called ceramide and is composed of a sphingosine and a fatty acid residue. Glycolipids (Figure 6.6) are classified into three groups according to the carbohydrate structure near the lipid–carbohydrate linkage, namely globo-series (Galα1 → 4Galβ1 → 4Glc), ganglio-series (GalNAcβ1 → 4Galβ1 → 4Glc) and neolacto-series (GlcNAcβ1 → 3Galβ1 → 4Glc). Furthermore, glycolipids with sialic acids are collectively called gangliosides.

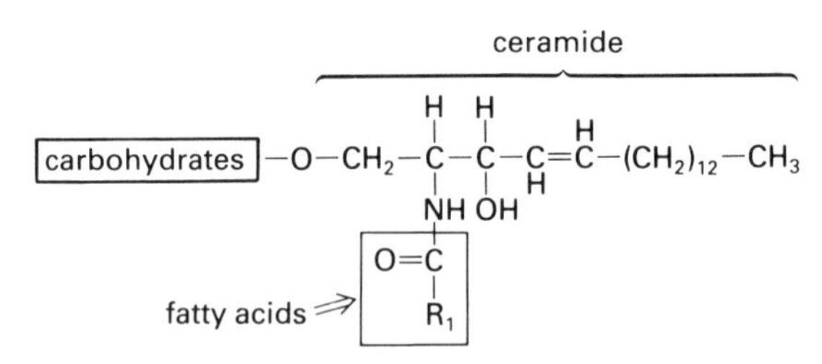

Figure 6.5 General structure of sphingoglycolipids.

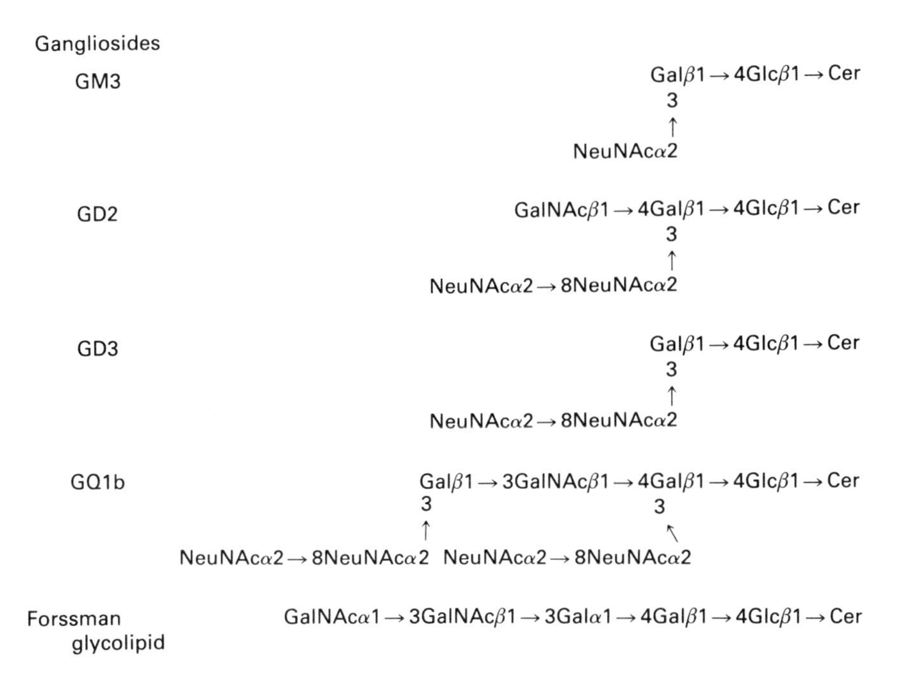

Figure 6.6 Examples of glycolipids.

Sugars at the distal end from the sugar involved in the carbohydrate–peptide or carbohydrate–lipid linkage are called non-reducing. Sugar structures at and near non-reducing ends are most important since they are critical in the recognition of carbohydrate sequences by most ligands.

6.1.2 Biosynthesis

Carbohydrates are basically formed by the sequential addition of monosaccharides from sugar nucleotides by glycosyltransferases (Figure 6.7). Therefore, the structure of carbohydrates is controlled by the specificity of glycosyltransferases involved in the synthesis. A large number of glycosyltransferases have been identified, corresponding to the diversity found in carbohydrate structures [6]. The core structure of asparagine-linked carbohydrates is synthesized in a form linked to dolichol phosphate. After *en bloc* transfer of a large oligosaccharide to the acceptor protein, the external portions of the precursor sugar chains are trimmed. The external sequences of the sugar chains are subsequently added by successive transfer

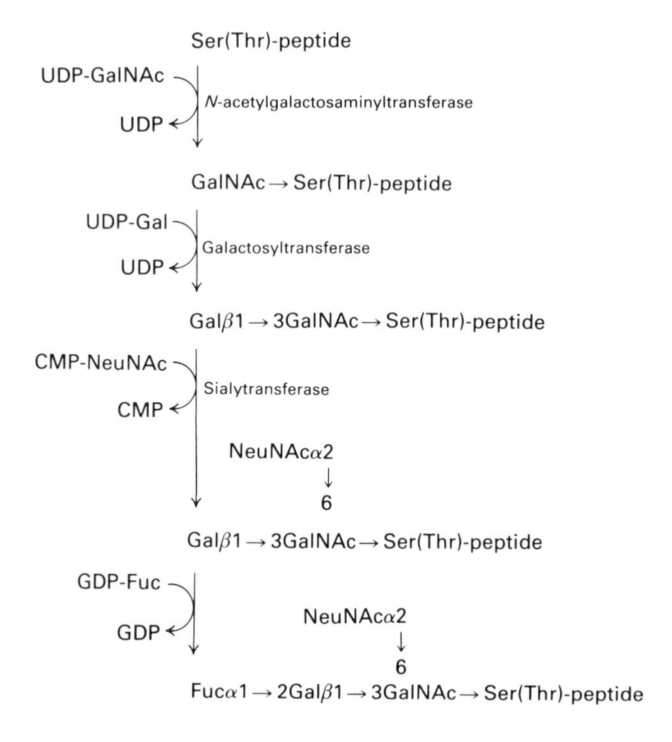

Figure 6.7 Step-wise action of different glycosyltransferases to form an oligosaccharide.

of sugar residues (Figure 6.8) [7]. Tunicamycin is an antibiotic, which inhibits the synthesis of lipid-linked oligosaccharides, while swainsonine and deoxynojirimycin inhibit the trimming process. They are widely used as reagents to inhibit the synthesis of specific carbohydrates, and to examine the effect of inhibition on a variety of cellular activities [8].

6.1.3 Recognition by antibodies, lectins, toxins and microorganisms

The structural differences between cell surface carbohydrates are often recognized by antibodies (Table 6.2). A typical example is ABH blood group antigens. Blood group A is determined by an *N*-acetylgalactosamine residue in a GalNAcα1 → 3(Fucα1 → 2)Gal sequence, and blood group B by a galactose residue in a Galα1→3(Fucα1→2)Gal sequence (Figure 6.9) [9]. Therefore, the presence or absence of an *N*-acetyl group in a specific position determines group A and B antigenicity.

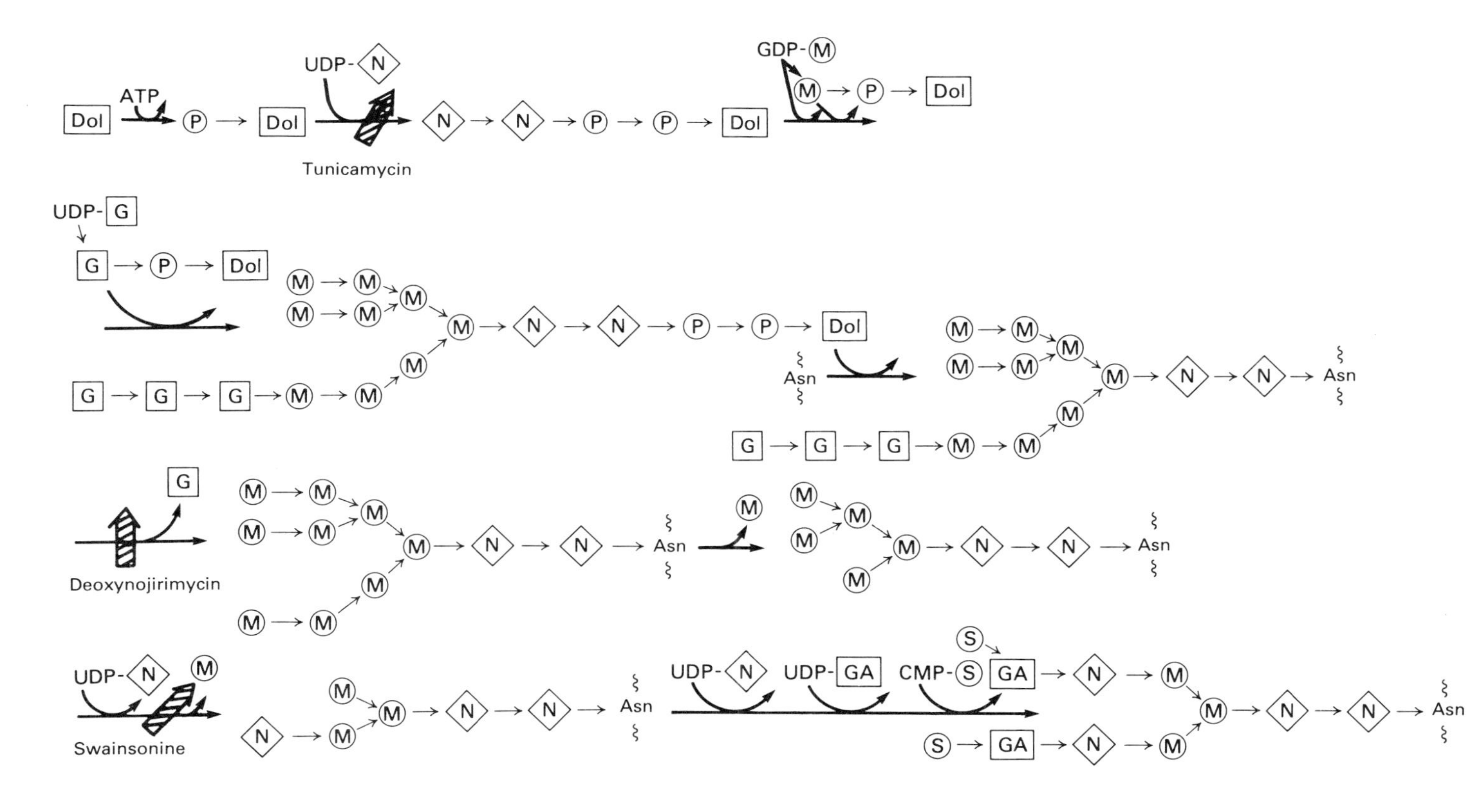

Figure 6.8 Formation of asparagine-linked oligosaccharide. Dol, dolichol (the lipid portion); P, phosphate; N, GlcNAc; M, mannose; G, glucose; GA, galactose; S, sialic acid.

Table 6.2 Examples of lectins and carbohydrate-recognizing monoclonal antibodies useful for cell identification and separation

Reagents	Specificity
Lectins	
Peanut agglutinin (PNA)	Galβ1 → 3GalNAc
Griffonia simplicifolia agglutinin I-B_4 (GSI-B_4)	Galα1 → 3Gal
Lotus tetragonolobus agglutinin (LTA or FBP)	Galβ1 → 4(Fucα1 → 3)GlcNAc, Fucα1 → 2Galβ1 → 4GlcNAc
Dolichos biflorus agglutinin (DBA)	GalNAc
Helix pomatia agglutinin (HPA)	GalNAc
Limulus polyphemus agglutinin (LPA)	sialic acid
Monoclonal antibodies	
SSEA-1	Galβ1 → 4(Fucα1 → 3)GlcNAc
SSEA-3	GalNAcβ1 → 3Galα1 → 4Gal
ECMA-2	Galα →
LD2	Galα1 → 3Gal
HNK-1	Sulphate-3-GlcAβ1 → 3Galβ1 → 4GlcNAc
	Galβ1 → 4GlcNAcβ1 → 3Galβ1 → 4GlcNAcβ1 → 3Galβ1 → 4GlcNAc

Lectins also detect structural differences in carbohydrates. Lectins are carbohydrate recognition proteins (distinct from antibodies) with cell agglutination activities (Table 6.2) [10]. Lectins have been isolated from diverse sources, notably plant seeds, and are widely used for cell identification and separation. Cell surface carbohydrates are also sites of attachment for pathogenic agents such as bacteria and viruses. For example, *Escherichia coli* has a lectin, which recognizes mannose, and uses the lectin for attachment to the surface of host cells [11]. Influenza virus binds to sialic acids on host cells [12]. Furthermore, certain bacterial toxins bind to cell surface carbohydrates; the receptor for *Cholerae* toxin is a ganglioside, GM1 [13].

6.1.4 Animal lectins

The specific recognition of cell surface carbohydrates by antibodies, lectins, toxins and microorganisms is probably not the major physiological function of the carbohydrate chains. However, such recognition gives us sufficient

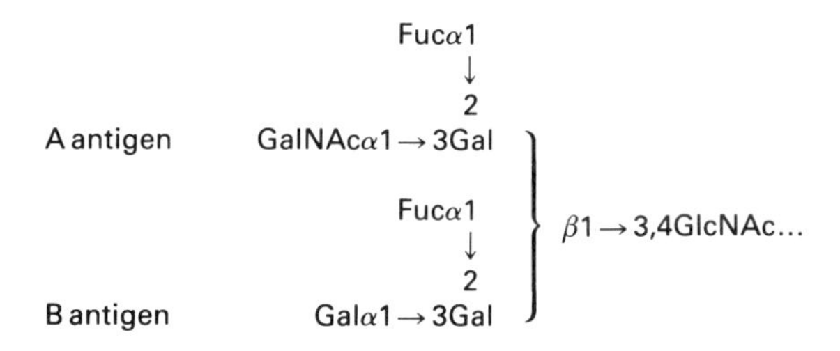

Figure 6.9 Structure of epitopes of blood group A and B antigens.

reason to believe that these structures are also specifically recognized by endogeneous proteins and are involved in many cellular processes. Indeed, lectins, especially β-galactoside-recognizing lectins, have been isolated from many vertebrate tissues. Notably, a hepatic β-galactoside-recognizing lectin has been identified as a receptor for endocytosing partly decomposed glycoproteins; it has been suggested that this lectin plays a critical role in removing these glycoproteins from the bloodstream [14, 15]. Furthermore, mannose-6-phosphate receptor is essential for the transportation of newly synthesized lysozomal enzymes to lysozome [16]. Animal lectins can be classified into C-type (Ca^{2+}-requiring type) and S-type (soluble type) according to molecular weights and requirement of Ca^{2+} or -SH reagent for activity. Hepatic β-galactoside binding lectin is the prototype of a C-type one. Common sequences have been found among C-type lectins, and S-type lectins also have distinct consensus sequences [17]. Intense studies are being carried out on the functions of these lectins, especially in relation to cell surface recognition.

6.2 GROWTH REGULATION

A breakthrough in work on the physiological function of cell surface carbohydrates came with the finding of transformation-dependent alterations of glycolipid profiles. In 1968, Hakomori found that polyoma virus-transformed BHK fibroblasts had less GM3 ganglioside than normal BHK cells [18]. In rapidly growing BHK cells, the glycolipid profile was similar to that of transformed cells. When cell growth was arrested by contact inhibition, the amount of GM3 increased. Similar phenomena have also been found in other fibroblast cell lines, although the species of glycolipids regulated by growth status are different. Hakomori proposed the term 'contact extension': in certain glycolipids, the terminal structure extends when cell growth is arrested by contact inhibition of cell growth, and this mechanism is destroyed by transformation. He has further proposed that these glycolipids may be involved in growth regulation [19]. Indeed, EGF-dependent growth of BHK cells is inhibited by adding GM3 into the culture medium. PDGF dependent growth of 3T3 cells is inhibited by GM1 and GM3. Growth of A431 epidermoid carcinoma cells is also inhibited by GM3. In these cells, GM3 has been found to inhibit receptor autophosphorylation of EGF receptor [20]. Lyso GM3, which is GM3 without fatty acids, is a more potent inhibitor of cell growth and of EGF receptor autophosphorylation. On the other hand, deacetylated GM3 promotes cell growth and enhances autophosphorylation of EGF receptor [21]. These results are consistent with the idea that activities of receptors for certain growth factors are modulated by glycolipids and their metabolites.

The pattern of asparagine-linked carbohydrates is also altered upon viral-transformation and upon changes of growth status. Thus, oligosaccharides with GlcNAcβ1 $\rightarrow$ 6Man branches increase in virally-transformed cells [22]. Furthermore, the ratio of high-mannose oligosaccharides to complex-type oligosaccharides increases in rapidly growing cells [23]. These general changes in cell surface can also influence the function of growth factor receptors. Increased GlcNAcβ1 $\rightarrow$ 6Man branching in mouse melanoma cells has been correlated with their increased metastatic potential [24].

6.3 MOUSE EMBRYOGENESIS

Cell surface carbohydrates undergo marked alterations during differentiation and development [25, 26, 27]. Some of the carbohydrate sequences preferentially expressed in restricted cell populations can be recognized by antibodies or lectins, and can be utilized for cell identification and separation (Table 6.2). Furthermore, some carbohydrate sequences preferentially expressed at certain stages of differentiation may be involved in the cell surface recognition necessary for differentiation and development. Detailed studies on carbohydrate changes during differentiation and development have been performed in early mouse embryogenesis, blood cell differentiation and neural differentiation.

Muramatsu *et al.* found glycoprotein-bound high-molecular-weight carbohydrates in EC cells and in preimplantation mouse embryos: EC cells cultured in the presence of radioactive fucose were extensively digested with pronase and the resulting glycopeptides were analysed by gel filtration on a column of Sephadex G-50. The glycopeptides from EC cells were found to have large amounts of high molecular weight species (molecular weight around 10000 or more (Figure 6.9) [28]. Fucose-labelled glycopeptides of such high molecular weight have not been detected in significant amounts in many normal and cancerous cells. Indeed, differentiated cells derived from EC cells scarcely had any of the large glycopeptides (Figure 6.10b,c).

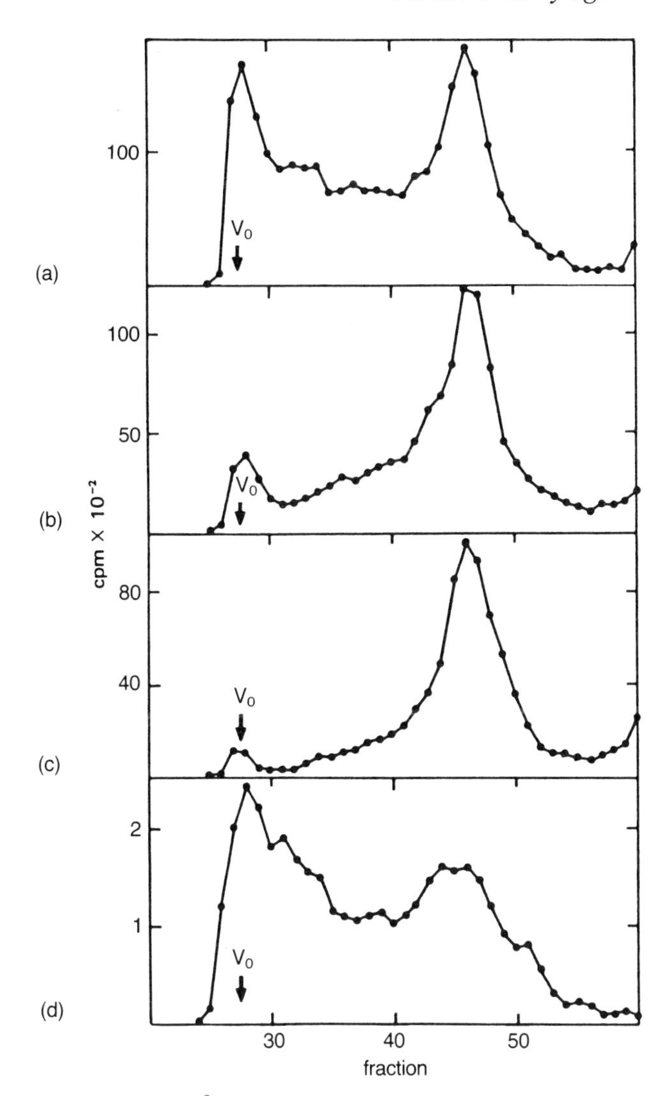

Figure 6.10 Analysis of [^{3}H]fucose-labelled glycopeptides by Sephadex G-50 column chromatography. (a) EC cells (PCC3 cells); (b) EC cells differentiated for 14 days; (c) EC cells differentiated for 28 days; (d) preimplantation mouse embryos; V_0, excluded volume. (Cited from [28].)

Preimplantation mouse embryos also synthesized the large glycopeptides (Figure 6.10d). The large carbohydrate called embryoglycan has been shown to belong to the poly-*N*-acetyllactosamines, which are carbohydrates with (Galβ1 $\rightarrow$ 4GlcNAcβ1 $\rightarrow$ 3)n repeating structures (Figure 6.11) [29, 30]. While poly-*N*-acetyllactosamines are present in diverse sources, only

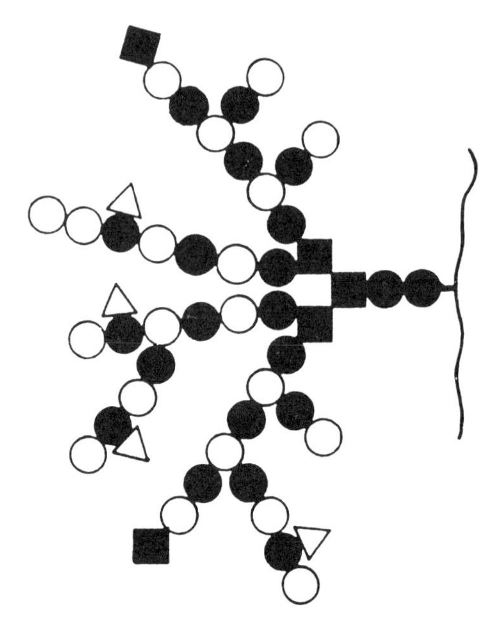

Figure 6.11 A model of a poly-*N*-acetyllactosamine from EC cells (embryoglycan). ○●■△, sugar residues.

those from EC cells, early embryonic cells and erythrocytes have structures complex enough to have a molecular weight of about 10 000 or more. Furthermore, embryoglycan and erythrocyte glycan can be distinguished by the cell surface markers they carry.

Embryoglycan has been shown to carry several cell surface carbohydrate markers preferentially expressed in early embryonic cells [25, 31–33]. During early embryogenesis, their mode of expression is altered in complicated ways (Table 6.3). Some of these markers are useful in the analysis of early embryogenesis and *in vitro* differentiation of EC cells. A notable example is SSEA-1, an antigen defined by a monoclonal antibody produced by Solter and Knowles [33]. SSEA-1 is expressed intensely in EC cells, but is absent in most cells differentiated from them. Therefore, loss of SSEA-1 can be used as an index of the differentiation of EC cells. On the other hand, *Dolichos biflorus* agglutinin (DBA) binding sites are expressed in extraembryonic endoderm cells and their precursors in early postimplantation embryos [34]. DBA binding sites are detected in some EC cells, such as F9 cells; these cells can be regarded as proceeding a step towards differentiation to extraembryonic endoderm cells. Indeed, F9 cells differentiate mostly to extraembryonic endoderm cells upon retinoic acid

Table 6.3 Altered expression of cell-surface markers during early stages of mouse embryogenesis

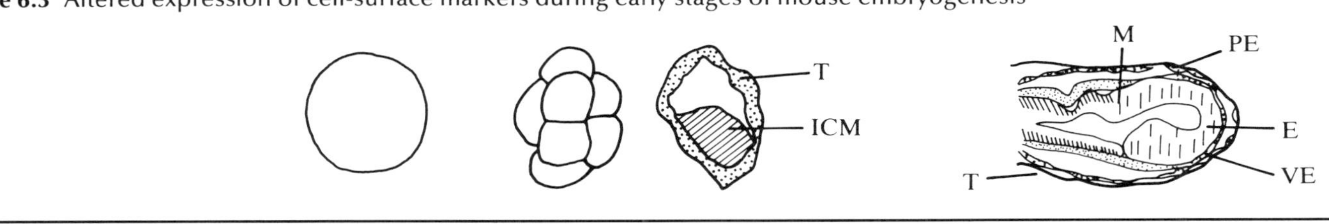

Name of reagents	Fertilized egg	8-cell stage	Blastocyst		Egg cylinder (6 day)				
			ICM	T	E	M	VE	PE	T
Markers on poly-*N*-acetyllactosamines									
SSEA-1, LTA	–	– → +	± → +	+	+	–	+	–	+
DBA	+	+	+ → ±	+ → –	–	–	+	+	–
GS-I	–	–	–	–	–	–	+	+	+
ECMA-2	– → +	+	±	+ → ±	±	–	+	–	– → +
i	–	–	–	–	–	–	+	+	–
Markers on globo-series glycolipids									
Forssman	–	–	+	+ → –	+ → –	–	+	+	–
SSEA-3	+	+	+	+ → –	–	–	+	–	–

ICM, inner cell mass; T, trophectoderm, trophoblastic giant cells; E, embryonic ectoderm; M, mesoderm; VE, visceral endoderm; PE, parietal endoderm.

treatment. In the case of multipotential EC cells, which do not express DBA binding sites, the binding sites serve as good markers of differentiation to extraembryonic endoderm cells and the precursors (Figure 6.12). In addition to markers on poly-*N*-acetyllactosamines, certain glycolipid markers also undergo alterations during early mouse embryogenesis (Table 6.3).

Several studies have been performed on the function of the poly-*N*-acetyllactosamines abundant in early embryonic cells. Shur fixed poly-*N*-acetyllactosamines from EC cells (embryoglycan) to a petri dish and found that EC cells became more adhesive to the modified dish. He has suggested that poly-*N*-acetyllactosamines are involved in the cell to cell adhesion of EC cells [35]. He has proposed that a galactosyltransferase is involved in recognizing poly-*N*-acetyllactosamines. $\beta 1 \rightarrow 4$galactosyltransferase is an enzyme which recognizes *N*-acetylglucosamine and transfers galactose from UDP-galactose to the acceptor. Indeed, the presence of galactosyltransferase on the surface of EC cells and early embryonic cells has been implicated by enzymological [36] and immunological evidence [37]. Therefore, it seems probable that the *N*-acetylglucosamine terminal on poly-*N*-acetyllactosamines is recognized by a galactosyltransferase on the surface of EC cells. As evidence that the

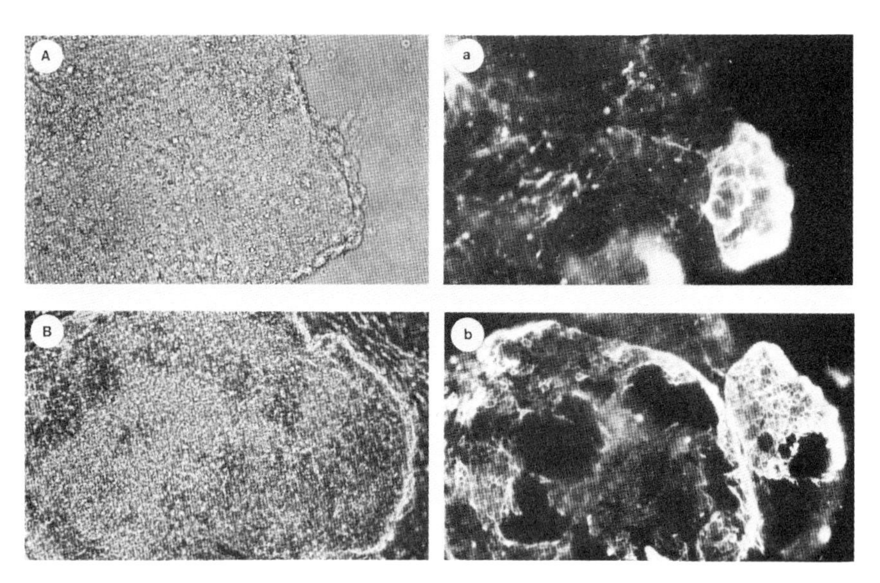

Figure 6.12 FITC-*Dolichos biflorus* agglutinin (DBA) reacts with extraembryonic endoderm cells (B, b) and the precursors (A, a) upon *in vitro* differentiation of EC cells. A, B, phase contrast; a, b, staining with FITC-DBA. (Cited from [84].)

recognition plays an important role, lactalbumin, a specific modifier of the galactosyltransferase, inhibits the adhesion of EC cells to the modified dish. Shur has further proposed that the recognition of poly-*N*-acetyllactosamines by the galactosyltransferase forms the basis of egg–sperm recognition [38] and also of the migration of embryonic cells.

Fenderson and Hakomori have added SSEA-1 determinant linked to a carrier to the culture medium of preimplantation embryos. The treated embryos cannot tightly associate, a process called compaction [39]. Feizi and co-workers have decompacted preimplantation embryos by removing Ca^{2+} and have treated them with endo-β-galactosidase which degrades poly-*N*-acetyllactosamines. The digested embryos are more difficult to re-compact than the undigested embryo [40]. Muramatsu and co-workers have observed that a monoclonal antibody reacting with a fucosyl residue in poly-*N*-acetyllactosamines inhibits cell–substratum adhesion of F9 EC cells (Figure 6.13) [41].

All these results indicate strongly that poly-*N*-acetyllactosamines on early embryonic cells are involved in cell surface recognition, especially in cell adhesion. Studies using mutant EC cells, however, do not lead to a unifying conclusion. A mutant EC cell lacking binding sites for peanut lectin (the marker is also carried by embryoglycan in EC cells) is impaired in cell-to-cell adhesion. Mutants lacking SSEA-1, or even whole embryoglycan, adhere to each other and differentiate normally [42]. This is probably because other surface molecules compensate for the loss of poly-*N*-acetyllactosamines in the mutant cells.

6.4 BLOOD CELL DIFFERENTIATION

Carbohydrates on the surface of blood cells have been the subject of extensive studies, and each type of blood cell is known to have unique sets of glycoconjugates. For example, comparing human erythrocytes and granulocytes, both of which have differentiated from common stem cells (myeloid stem cells), the following differences have been found in the structures of cell surface carbohydrates:

1. Poly-*N*-acetyllactosamines from erythrocytes have high molecular weights (around 10 000), are highly branched and carry determinants of ABH blood group antigens [43, 44]. On the other hand, poly-*N*-acetyllactosamines from granulocytes are of medium molecular weight (around 5 000), have linear poly-*N*-acetyllactosamine chains and do not carry ABH blood group antigen but carry SSEA-1 determinant [45].
2. Erythrocytes have short oligosaccharides *O*-glycosidically linked to the serine or threonine residues of proteins such as glycophorin, and the major oligosaccharide in humans is NeuNAcα2$\rightarrow$3Galβ1$\rightarrow$

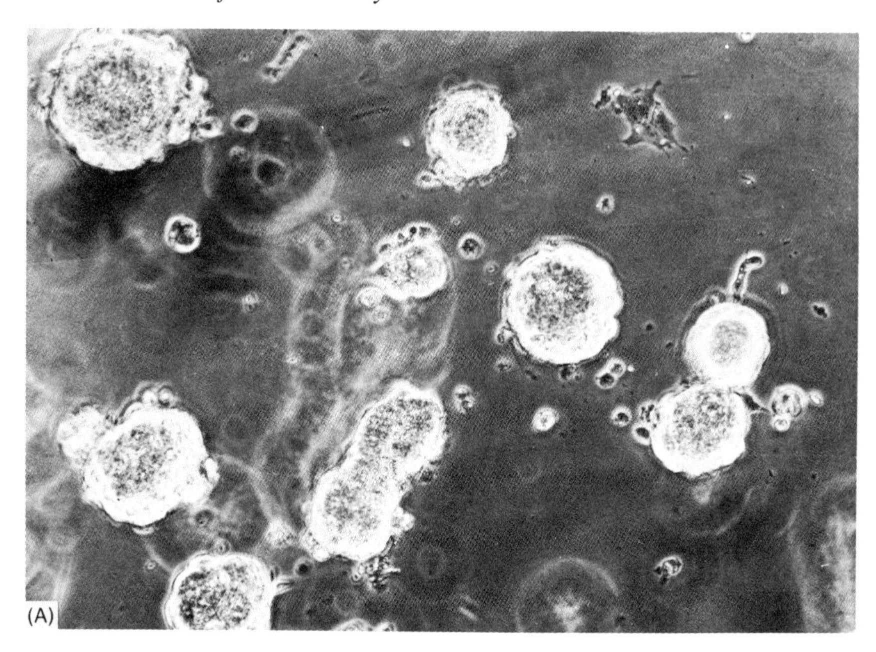

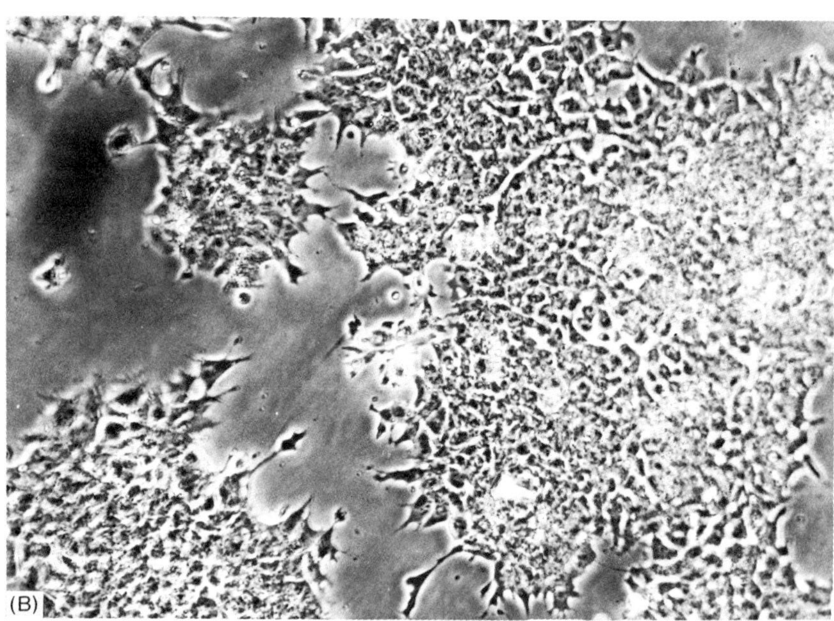

Figure 6.13 Monoclonal antibody reacting with fucosyl poly-*N*-acetyllactosamines inhibits cell–substratum adhesion of F9 embryonal carcinoma cells. Cells are cultured with an antibody reacting with fucosyl poly-*N*-acetyllactosamine (a) or with Forssman antigen (b). (Reprinted from [41].) ©Academic Press.

3(NeuNAcα2$\rightarrow$6)GalNAc [46]. Oligosaccharides *O*-glycosidically linked to serine and threonine residues in granulocytes scarcely have any NeuNAcα2 $\rightarrow$ 6GalNAc structures but have GlcNAcβ1$\rightarrow$6GalNAc structures [47].

3. Globoside is the major glycolipid in erythrocytes. Granulocytes do not have significant amounts of globoside, but have complex glycolipids of the general structure

$$[\text{Gal}\beta1 \rightarrow 4(\pm\text{Fuc}\alpha1 \rightarrow 3)\text{GlcNAc}\beta1 \rightarrow 3)]\text{nGal}\beta1 \rightarrow 4\text{Glc}\beta \rightarrow \text{Cer}$$

where $n = 0 - 3$ [48].

Although the poly-*N*-acetyllactosamines of erythrocytes of ordinary adults are branched, those in the fetus are linear; during human embryogenesis, the basic structure of poly-*N*-acetyllactosamine in the erythrocytes is altered [49].

Carbohydrate structures differ between lymphocyte subpopulations [50, 51]. Thus, *Vicia villosa* lectin, peanut agglutinin and *Limulus polyphemus* agglutinin are useful in the isolation of cytotoxic T cells, suppressor T cells and helper T cells, respectively [52, 53]. Among these markers, binding sites for *Vicia villosa* lectin are glycoproteins of 145 K, and the lectin binding site is an *O*-glycosidically linked oligosaccharide of GalNAcβ1$\rightarrow$4Gal structure [54].

In several human haematopoietic cell lines, sialylated *O*-glycosidically-linked oligosaccharides are mainly found on a glycoprotein called leukosialin. From cDNA cloning and sequencing, this molecule has been shown to have 371 amino acids, of which 224 residues are outside the cell [55]. One out of 3 extracellular amino acids appears to be *O*-glycosylated. Fukuda and co-workers have found that *O*-glycosidically-linked oligosaccharides of leukosialin are different in different cell lines [56]. Thus, in K562 erythroid cells, the major oligosaccharides in the glycoprotein are NeuNAcα2$\rightarrow$3Galβ1$\rightarrow$3GalNAc, Galβ1$\rightarrow$3(NeuNAcα2$\rightarrow$6)GalNAc and NeuNAcα2$\rightarrow$3Galβ1$\rightarrow$3(NeuNAcα2$\rightarrow$6)GalNAc. In HL-60 promyelocytic leukemia cells and HSB-2 T-lymphoid cells, the major oligosaccharide in the glycoprotein is NeuNAcα2$\rightarrow$3Galβ1$\rightarrow$4GlcNAcβ1$\rightarrow$6(NeuNAcα2$\rightarrow$3Galβ1$\rightarrow$3)GalNAc. As pointed out in the comparison of carbohydrates from erythrocytes and granulocytes, formation of the two structures, NeuNAcα2$\rightarrow$6GalNAc and GlcNAcβ1$\rightarrow$6GalNAc appear to be important steps which are developmentally regulated in the differentiation of blood cells (Figure 6.14).

When HL-60 promyelocytic leukemia cells are treated with TPA, they differentiate to macrophages. Accompanying the differentiation to macrophages, a glycolipid GM3 has been found to increase. Saito and co-workers have added GM3 to culture medium of HL-60 cells and have been able to induce the differentiation of the cells (Figure 6.15) [57]. Therefore,

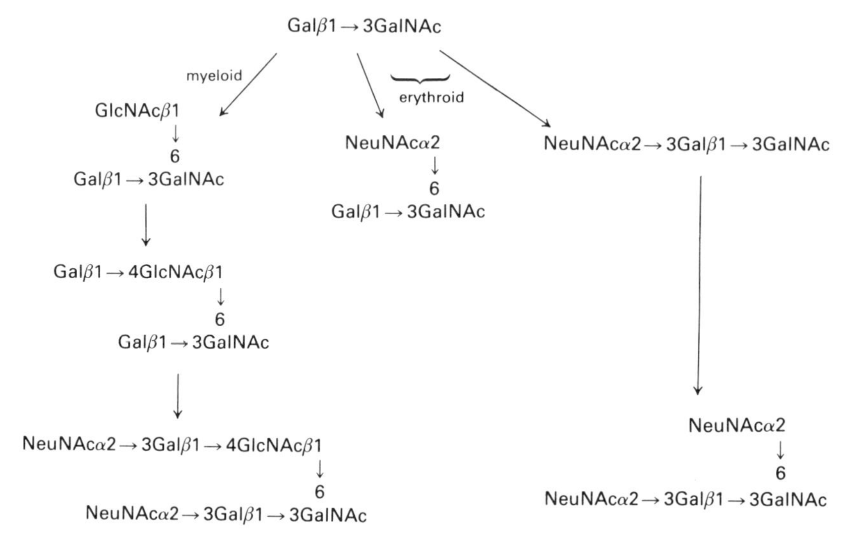

Figure 6.14 A different pathway of biosynthesis of oligosaccharides in myeloid (left) and erythroid (right) cell lineages. (Cited from [47].)

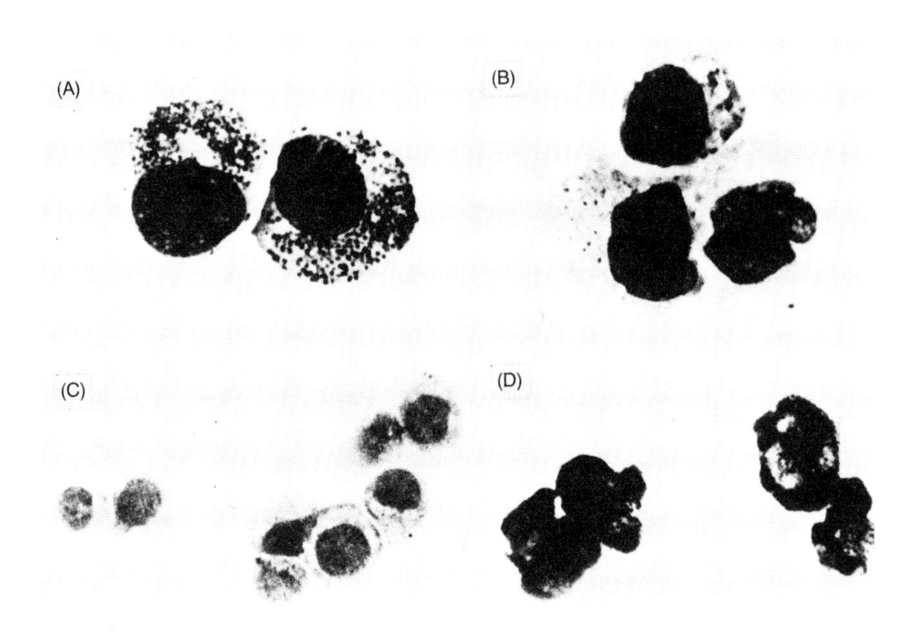

Figure 6.15 GM3 induces HL-60 leukemia cells to differentiate into macrophages. (a), (c) Undifferentiated cells; (b), (d) cells differentiated by GM3. (a), (b) Wright–Giemsa staining; (b), (d) histochemical demonstration of α-naphthyl butyrate esterase, a marker enzyme of macrophages. (Reprinted from [57].)

increase of GM3 may be a causative factor triggering the differentiation of these cells. Furthermore, complex glycolipids with Galβ1 $\rightarrow$ 4GlcNAc repeating units increase upon differentiation of HL-60 cells to granulocytes by treatment with dimethylsulphoxide. When the glycolipid is added to HL-60 cells, the cells differentiate to granulocytes [58]. These results suggest that glycolipids may even be involved in the selection of the direction of differentiation.

6.5 NERVE CELL DIFFERENTIATION

Nerve cells are rich in glycoconjugates, and they are believed to be involved in the development of nervous systems, especially formation of the neural network. Furthermore, certain carbohydrates have been useful in the identification of cell populations in nervous tissues undergoing differentiation. For example, galactocerebroside has been used as a marker for oligodendrocytes, while A2B5 antigen, a kind of ganglioside antigen distinguishes two types of astrocytes (Chapter 4).

Finne found asparagine-linked oligosaccharides with polysialyl units [(NeuNAcα2$\rightarrow$8NeuNAc)$n\alpha$2$\rightarrow$3Gal] in the developing rat brain; this carbohydrate was absent in the adult brain [59]. N-CAM has been identified as the major glycoprotein carrying the polysialyl units [60]. N-CAM isolated from the adult brain has apparent molecular weight 110 K to 160 K. The glycoprotein isolated from the embryonic brain has molecular weight 200 K to 250 K. This difference is due to the presence of polysialo-units in embryonic N-CAM. Edelman and co-workers have found that the polysialo-carbohydrate weakens homophilic interaction of N-CAM; as a result, nerve cells can move more freely in the embryonic period [61]. In some mutant mice a defect may be present in the synthesis of the glycan and cause poor movement of nerve cells, leading to aberrant construction of the nerve network. Rutishauser *et al.* have used endo-neuraminidase from bacteriophage KIF, which hydrolyzes polysialo-carbohydrates [62]. When the enzyme is added to a culture of dorsal root ganglia, the neurite outgrowth becomes thicker and the number of outgrowths decreases. This phenomenon can be interpreted as increased N-CAM activity resulting from the removal of polysialo-units. Furthermore, when the enzyme is injected to the eye of 3.5 day chick embryos, the treated eyes change markedly in shape. Especially noticeable is thickening of the neural retina in the dorsal-posterior region.

HNK-1 antigen is defined by an unusual carbohydrate epitope, sulphate-3-glucuronic acid-β1 $\rightarrow$ 3Galβ1 $\rightarrow$ 4GlcNAc [63]. The antigen was initially found by Ilyas *et al.* using IgM paraproteins from patients with autoimmune neuropathy. This antigen is predominantly expressed in the embryonic period, especially in post-migratory cells [64]. The antigenic epitope has been found both in glycoproteins and glycolipids. Glycoproteins carrying

HNK-1 antigen are so far all involved in cell adhesion; they are N-CAM, L1,J1 and myelin-associated glycoprotein. When a monoclonal antibody, which recognizes HNK-1 antigen, is added to the culture of nerve cells, neuron–astrocyte and astrocyte–astrocyte adhesion is specifically inhibited [65]. Furthermore, an oligosaccharide with HNK-1 determinant inhibits some of the adhesion phenomena. Therefore, HNK-1 antigen may be involved in the adhesion of nerve cells.

Neural tissues are rich in gangliosides. Nagai and co-workers have found that a specific ganglioside, GQ1b, promotes the neurite extension of cultured neuroblastoma cells [66]. Other gangliosides are ineffective.

An antigen with Le^x epitope shows developmentally regulated expression during brain development. In the mouse, the antigen becomes detectable in the cerebral cortex on day 11 of embryogenesis [67]. On day 13–15, ventricular and subventricular zones of the cortex are strongly antigen-positive. By birth the antigen has disappeared. Even in the adult, carbohydrate antigens are differentially expressed in different neurons. Subsets of dorsal root ganglion neurons of the rat can be distinguished by globo-series markers (SSEA-3 and SSEA-4) and by neo-lacto-series markers (Galβ1 $\rightarrow$ 4GlcNAc and Galα1 $\rightarrow$ 3Galβ1 $\rightarrow$ 4GlcNAc). The areas where a β-galactoside binding lectin is expressed overlaps generally with the area where the neo-lacto-series markers are expressed [68].

6.6 OTHER SYSTEMS

In kidney morphogenesis, the nephrogenic mesenchyme and the ureter epithelium interact, and as a result the epithelium branches and the mesenchyme converts into a new epithelium. Using explants of the embryonic kidney, this developmental process can be performed *in vitro*. Sariola *et al.* have found that a monoclonal antibody reacting with GD3 inhibits both epithelium branching and mesenchyme differentiation *in vitro* [69]. Thus, GD3 ganglioside appears to play a role in the epithelial–mesenchymal interaction.

Even in adult tissues, cells differentiate from the corresponding stem cells, and carbohydrate alterations have been noticed during the process of differentiation; i.e. differentiation in the skin [70] in the epithelium of the oesophagus [71] and in spermatogenesis [72]. In the last case, the change observed is spermatogonium [PNA(−), SBA(−)] → primary spermatocyte [PNA(+), SBA(−)] → secondary spermatocyte [PNA(+), SBA(+)]. Such knowledge will be helpful in isolating some intermediate cells of the differentiation process. The physiological role of the carbohydrate alterations on these occasions is not known, while it is interesting to speculate that the stage-specific carbohydrates are recognized by endogeneous lectins and are involved in keeping cells at the same differentiation stage in a cell layer.

6.7 ROLES IN CELL ADHESION

The role of cell surface carbohydrates in cell adhesion is becoming increasingly clear. In addition to the evidence reviewed above, the following results strengthen such a role. Cheresh *et al.* have found that a monoclonal antibody reacting with ganglioside GD2 and GD3 inhibits cell–substratum adhesion of cultured human melanoma cells. Vitronectin receptor, which is a member of the integrin family (Chapter 5) has been isolated from the cell and found to contain tightly bound GD2 [73]. When the receptor is treated with EDTA, calcium ions are removed and the receptor activity is lost. Concomitantly GD2 dissociates from the receptor. Addition of Ca^{2+} to the EDTA-treated receptor restores its activity, and further addition of GD2 slightly but significantly increases the activity. Vitronectin receptor probably has a hole in which GD2 is retained; the glycolipid in turn appears to alter the three dimensional structure of the receptor and modulates its activity.

Laminin can agglutinate trypsinized, glutaraldehyde-fixed rabbit erythrocytes (Figure 6.16), and the agglutination reaction is inhibited by porcine gastric mucin and gangliosides. Thus, laminin can be considered to

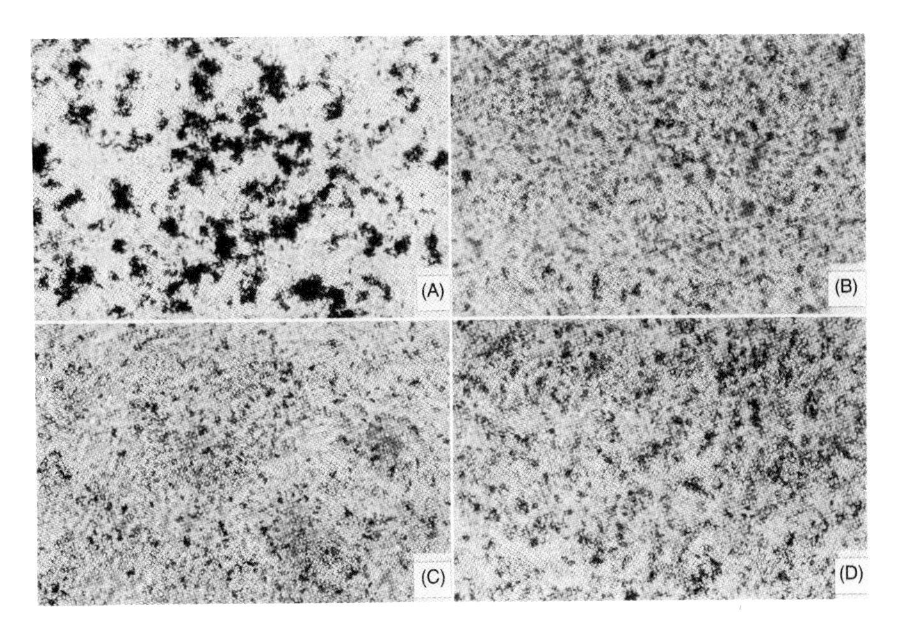

Figure 6.16 Lectin activity of laminin (a) and its inhibition by EGTA (b), EDTA (c) and low temperature (4°C) (d). Laminin (7.5 μg) was mixed with trypsinized, glutaraldehyde-fixed rabbit erythrocytes in 0.1 ml of reaction mixture. (Cited from [74].)

be a lectin [74]. To elucidate the carbohydrate structure recognized by laminin, Ginsburg and co-workers have examined various glycolipids fixed on thin layer plates for the binding activity to laminin. Sulphoglycolipids such as galactose-3-sulphate ceramide has been found to be strongly reactive. Therefore, laminin has a domain reacting with sulphated galactosyl residues, and this recognition should be important in the interaction of laminin with cell surfaces [75]. Thrombospondin and von Willebrand factor, two other adhesive proteins, also react with sulphoglycolipids. Recognition of sulphated galactosyl residues may have a broad significance in cell surface recognition.

Cellular slime molds grow as an amoeba feeding on bacteria. Upon starvation, they form a slug-like aggregate. The aggregate migrates and then differentiates to a structure called a fruiting body. The anterior portion of it becomes a head which will have many spores, and the posterial portion becomes a stalk. Glycopeptides from early stages of these aggregates inhibit aggregation of the cells at the same stage, but not the aggregation of cells at later stages [76].

Carbohydrate–protein recognition also appears to be of critical importance in fertilization. Mammalian eggs are surrounded by an external layer called Zona pellucidae. Three glycoproteins are identified in this layer of the mouse egg, and one called Zp3 is known to bind with sperm. When Zp3 is treated with mild alkali, *O*-glycosidically-linked oligosaccharides are released from the protein. One of the oligosaccharides has been found to have the ability to bind to sperm. Digestion with α-galactosidase inactivates the sperm binding activity of the oligosaccharide [77]. Therefore, the oligosaccharide moiety with an α-galactosyl terminus is likely to be important in sperm–egg binding.

6.8 COMMENTS

Cell surface carbohydrates are involved in many physiological processes. Their functions may be classified as *trans* (mediator or direct) ones and *cis* (modulator or indirect) ones (Figure 6.17). In the *trans* function, carbohydrates are recognized by carbohydrate-recognizing proteins, and the recognition alters the three dimensional structure of the carbohydrate-recognizing proteins or the carbohydrate-containing proteins. The structural change is transduced into the cell, or the carbohydrate–ligand interaction is important in cell adhesion. This mode of *trans* function is expected to occur widely, since lectins are broadly distributed in animal tissue. Furthermore, a carbohydrate-binding domain is present in seemingly unrelated molecules, such as core proteins of proteoglycans [78].

In the *cis* function, carbohydrates modulate the function of proteins, especially transmembrane proteins, to which the carbohydrates are

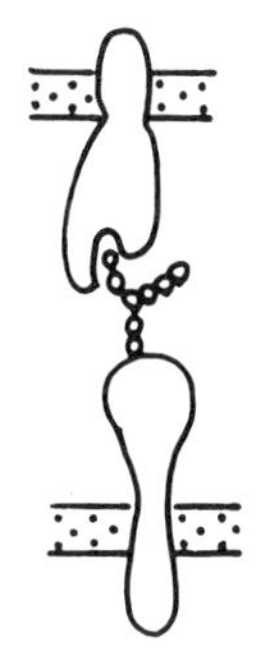

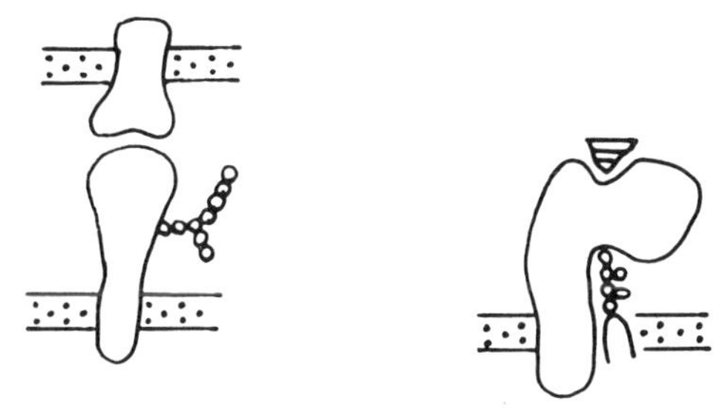

Figure 6.17 Roles of cell surface carbohydrates in cell surface recognition.

covalently or non-covalently associated. An example is the negative modulation of N-CAM function by polysialo-carbohydrate. Requirement of asparagine-linked oligosaccharides for efficient recognition of the Fc region of IgG by Fc receptor and by complement has been well documented [79, 80]. Considering that modulating oligosaccharides are linked to immunoglobulin (-like) domains both in IgG and N-CAM, it is predicted that in many members of the immunoglobulin superfamily, oligosaccharides are performing a *cis*-function. Another example of a *cis* function is the positive modulation of vitronectin receptor by GD2. Furthermore, modulation of autophosphorylation of growth factor receptors by gangliosides is probably a highly significant observation. Glycolipids might perform many of the functions postulated, i.e. growth-regulation, induction of differentiation, neurite extension and epithelial–mesenchyme interaction by modulating the activity of cell surface receptors, in particular those of growth factors. In addition to protein–carbohydrate interactions, carbohydrate–carbohydrate interactions may be also important in cell surface recognition. Hakomori and co-workers have found recently that two SSEA-1 structures are complementary and have proposed that the SSEA-1–SSEA-1 interactions play a role in adhesion of early embryonic cells [81].

So far the functions of cell surface carbohydrates have been studied by biochemical, immunological and cell biological approaches. Molecular biological approaches will soon be introduced to the field and are expected to create a great deal of excitement. cDNA and genomic clones of glycosyltransferases, lectins and core proteins of glycoproteins have been isolated in certain cases, and many more of them will be isolated in a few years. After changing the control region of the genes to force inappropriate expression, one can transfect the altered gene into cells with differentiating capacity or produce transgenic animals harbouring the altered gene. Alterations in differentiation and development caused by such manipulations will not only confirm many of the functional roles of cell surface carbohydrates proposed on the basis of indirect evidence, but also reveal their hitherto unexpected functions. Advances in homologous recombination techniques, which can produce defective mutants, are also badly needed in this field.

REFERENCES

1. Hughes, R.C. (1983) *Glycoproteins*, Chapman and Hall, London UK
2. Sharon, N. and Lis, H. (1982) In *The Proteins, Vol. V* (eds H. Neurath and R.L. Hill), Academic Press, New York, USA, pp. 1–144
3. Montreuil, J., Bouquelet, S., Debray, H., Fournet, B., Spik, G. and Strecker, G. (1986) *Carbohydrate Analysis, A Practical Approach* (eds M.F. Chaplin and J.F. Kennedy) IRL Press, Oxford, UK, pp. 43–204
4. Hakomori, S. (1981) *Ann. Rev. Biochem.*, **50**, 733–4
5. Holt, G.D. and Hart, G.W. (1986) *J. Biol. Chem.*, **261**, 8049–57
6. Beyer, T.A., Sadler, J.E., Rearick, J.I., Paulson, J.C. and Hill, R.L. (1981) *Adv. Enzymol.*, **52**, 23–175
7. Kornfeld, S., Li, E. and Tabas, I. (1978) *J. Biol. Chem.*, **253**, 7771–8
8. Murray, R.K., Granner, D.K., Mayes, P.A. and Rodwell, V.W. (1988) *Harper's Biochemistry*, Appleton and Lange, Norwalk, USA
9. Watkins, W.M. (1972) In *Glycoproteins* (ed. A. Gottschalk) Elsevier, Amsterdam, Netherlands, pp. 830–91
10. Liener, I.E., Sharon, N. and Goldstein, I.J. (eds) (1986) The Lectins, Academic Press, Orlando, USA
11. Ofek, I., Mirelman, D. and Sharon, N. (1977) *Nature*, **265**, 623–5
12. Suzuki, Y., Matsunaga, M. and Matsumoto, M. (1985) *J. Biol. Chem.*, **260**, 1362–5
13. Cuatracasas, P. (1973) *Biochemistry*, **12**, 3547–58
14. Ashwell, G. and Morell, A.G. (1974) *Advan. Enzymol.*, **41**, 99–128
15. Hudgin, R.L., Pricer Jr., W.E., Ashwell, G., Stockert, R.J. and Morell, A.G. (1974) *J. Biol. Chem.*, **249**, 5536–43

16. Kaplan, A., Achord, D.T. and Sly, W.S. (1977) *Proc. Natl. Acad. Sci.*, **74**, 2026–30
17. Drickamer, K. (1988) *J. Biol. Chem.*, **263**, 9557–60
18. Hakomori, S. and Murakami, W.T. (1968) *Proc. Natl. Acad. Sci.*, **59**, 254–61
19. Hakomori, S. (1975) *Biochim. Biophys. Acta.*, **417**, 55–89
20. Bremer, E.G., Schlessinger, J. and Hakomori, S. (1986) *J. Biol. Chem.*, **261**, 2434–40
21. Hanai, N., Dohi, T., Nores, G.A. and Hakomori, S. (1988) *J. Biol. Chem.*, **263**, 6296–301
22. Yamashita, K., Ohkura, T., Tachibana, Y., Takasaki, S. and Kobata, A. (1984) *J. Biol. Chem.*, **259**, 10834–40
23. Muramatsu, T., Koide, N., Ceccarini, C. and Atkinson, P.H. (1976) *J. Biol. Chem.*, **251**, 4674–9
24. Dennis, J.W., Lajerte, S., Waghorne, C., Breitman, M.L. and Kerbel, R.S. (1987) *Science*, **236**, 582–5
25. Muramatsu, T. (1988) *J. Cell. Biochem.*, **36**, 1–14
26. Muramatsu, T. (1988) *Biochimie*, **70**, 1589–96
27. Feizi, T. (1985) *Nature*, **314**, 53–7
28. Muramatsu, T., Gachelin, G., Nicolas, J.F., Condamine, H., Jakob, H. and Jacob, F. (1978) *Proc. Natl. Acad. Sci.*, **75**, 2315–9
29. Muramatsu, H., Ishihara, H., Miyauchi, T., Gachelin, G., Fujisaki, T. *et al.* (1983) *J. Biochem.*, **94**, 799–810
30. Kamada, Y., Arita, Y., Ogata, S., Muramatsu, H. and Muramatsu, T. (1987) *Eur. J. Biochem.*, **163**, 497–502
31. Muramatsu, T., Gachelin, G., Damonneville, M., Delarbre, C. and Jacob, F. (1979) *Cell*, **18**, 183–91
32. Ozawa, M., Muramatsu, T. and Solter, D. (1985) *Cell Differ.*, **16**, 169–73
33. Solter, D. and Knowles, B.B. (1978) *Proc. Natl. Acad. Sci.*, **75**, 5565–9
34. Noguchi, M., Noguchi, T., Watanabe, M. and Muramatsu, T. (1982) *J. Embryol. Exp. Morphol.*, **72**, 39–52
35. Shur, B.D. (1983) *Dev. Biol.*, **99**, 360–72
36. Shur, B.D. (1982) *J. Biol. Chem.*, **257**, 6871–8
37. Muramatsu, T. (1989), in *Carbohydrate Recognition in Cellular Function*, Ciba Foundation Symposium 145, pp. 273–4
38. Lopez, L.C., Bayna, E.M., Litoff, D., Shaper, N.L., Shaper, J.H. and Shur, B.D. (1985) *J. Cell Biol.*, **101**, 1501–10
39. Fenderson, B.A., Zahavi, U. and Hakomori, S. (1984) *J. Exp. Med.*, **160**, 1591–6
40. Rastan, S., Thorpe, S.J., Scudder, P., Brown, S., Gooi, H.C. and Feizi, T. (1985) *J. Embryol. Exp. Morphol.*, **87**, 115–28
41. Nomoto, S., Muramatsu, H., Ozawa, M., Suganuma, T., Tashiro, M. and Muramatsu, T. (1986) *Exp. Cell Res.*, **164**, 49–62
42. Dráber, P. and Malý, P (1987) *Proc. Natl. Acad. Sci.*, **84**, 5798–802
43. Järnefelt, J., Rush, J., Li, Y.-T. and Laine, R.A. (1978) *J. Biol. Chem.*, **253**, 8006–9
44. Krusius, T., Finne, J. and Rauvala, H. (1978) *Eur. J. Biochem.*, **92**, 289–300
45. Spooncer, E., Fukuda, M., Klock, J.C., Oates, J.E. and Dell, A. (1984) *J. Biol. Chem.*, **259**, 4792–801
46. Thomas, D.B. and Winzler, R.J. (1969) *J. Biol. Chem.*, **244**, 5943–6
47. Fukuda, M., Carlsson, S.R., Klock, J.C. and Dell, A. (1986) *J. Biol. Chem.*, **261**, 12796–806

48. Fukuda, M.N., Dell, A., Oates, J.E., Wu, P., Klock, J.C. and Fukuda, M. (1985) *J. Biol. Chem.*, **260**, 1067–82
49. Fukuda, M., Fukuda, M.N. and Hakomori, S. (1979) *J. Biol. Chem.*, **254**, 3700–03
50. Sharon, N. (1983) *Advan. Immunol.*, **34**, 213–98
51. Reisner, Y., Linker-Israeli, M. and Sharon, N. (1976) *Cell Immunol.*, **25**, 129–34
52. Kimura, A.K., Wigzell, H., Holmquist, G., Erisson, B. and Carlson, P. (1978) *J. Exp. Med.*, **149**, 473–84
53. Nakano, T., Imai, Y., Naiki, M. and Osawa, T. (1980) *J. Immunol.*, **125**, 1928–32
54. Conzelmann, A. and Kornfeld, S. (1984) *J. Biol. Chem.*, **259**, 12536–42
55. Killeen, N., Barclay, A.N., Willis, A.C. and Williams, A.F. (1987) *EMBO J.*, **6**, 4029–34
56. Carlsson, S.R., Sasaki, H. and Fukuda, M. (1986) *J. Biol. Chem.*, **261**, 12787–95
57. Nojiri, H., Takaku, F., Terui, Y., Miura, Y. and Saito, M. (1986) *Proc. Natl. Acad. Sci.*, **83**, 782–6
58. Nojiri, H., Kitagawa, S., Nakamura, M., Kirito, K., Enomoto, Y. and Saito, M. (1988) *J. Biol. Chem.*, **263**, 7443–6
59. Kinne, J. (1982) *J. Biol. Chem.*, **257**, 11966–70
60. Finne, J., Finne, U., Deagostini-Bazin, H. and Coridis, C. (1983) *Biochem. Biophys. Res. Commun.*, **112**, 482–7
61. Hoffman, S. and Edelman, G.M. (1983) *Proc. Natl. Acad. Sci.*, **80**, 5762–6
62. Rutishauser, U., Watanabe, M., Silver, J., Troy, F.A. and Vimr, E.R. (1985) *J. Cell Biol.*, **101**, 1842–9
63. Chou, D.K.H., Ilyas, A.A., Evans, J.E., Costello, C., Quarles, R.H. and Jungalwala, F.B. (1986) *J. Biol. Chem.*, **261**, 11717–25
64. Schwarting, G.A., Jungalwala, F.B., Chou, D.K.H., Boyer, A.M. and Yamamoto, M. (1987) *Dev. Biol.*, **120**, 65–76
65. Keilhauer, G., Faissner, A. and Schachner, M. (1985) *Nature*, **316**, 728–30
66. Tsuji, S., Arita, M. and Nagai, Y. (1983) *J. Biochem.*, **94**, 303–06
67. Yamamoto, M., Boyer, A.M., Schwarting, G.A. (1985) *Proc. Natl. Acad. Sci.*, **82**, 3045–9
68. Regan, L.J., Dodd, J., Bahrondes, S.H., Jessell, T.M. (1986) *Proc. Natl. Acad. Sci.*, **83**, 2248–52
69. Sariola, H., Aufderheide, E., Bernhard, H., Henke-Fahle, S., Dippold, W. and Ekbolm, P. (1988) *Cell*, **54**, 235–45
70. Brabec, R.K., Peters, B.P., Bernstein, I.A., Gray, R.H. and Goldstein, I.J. (1980) *Proc. Natl. Acad. Sci.*, **77**, 477–9
71. Watanabe, M., Muramatsu, T., Shirane, H. and Ugai, K. (1981) *J. Histochem. Cytochem.*, **29**, 779–90
72. Watanabe, M., Takeda, Z., Urano, H. and Muramatsu, T. (1982) In *Teratocarcinoma and Embryonic Cell Interactions* (eds T. Muramatsu *et al.*) Japan Scientific Societies Press/Academic Press, Tokyo, Japan, pp. 217–28
73. Cheresh, D.A., Pytela, R., Pierschbacher, M.D., Klier, F.G., Ruoslahti, E. and Reisfeld, R.A. (1987) *J. Cell Biol.*, **105**, 1163–73
74. Ozawa, M. and Muramatsu, T. (1983) *J. Biochem.*, **94**, 479–85
75. Roberts, D.D., Rao, C.N., Liotta, L.A., Gralnick, H.R. and Ginsburg, V. (1986) *J. Biol. Chem.*, **261**, 6872–7
76. Ziska, S.E. and Henderson, E.J. (1988) *Proc. Natl. Acad. Sci.*, **85**, 817–21
77. Wassarman, P.M. and Bleil, J.D. (1988) *Proc. Natl. Acad. Sci.*, **85**, 6778–82

78. Sai, S., Tanaka, T., Kosher, R.A. and Tanzer, M.L. (1986) *Proc. Natl. Acad. Sci.*, **83**, 5081–5
79. Koide, N., Nose, M. and Muramatsu, T. (1977) *Biochem. Biophys. Res. Commun.*, **75**, 838–44
80. Nose, M. and Wigzell, H. (1983) *Proc. Natl. Acad. Sci.*, **80**, 6632–6
81. Eggens, I., Fenderson, B., Toyokuni, T., Dean, B., Stroud, M. and Hakomori, S. (1989) *J. Biol. Chem.*, **264**, 9476–84
82. Fukuda, M. and Hakomori, S. (1979) *J. Biol. Chem.*, **254**, 5451–7
83. Yoshima, H., Furthmayr, H. and Kobata, A. (1980) *J. Biol. Chem.*, **255**, 9713–8
84. Muramatsu, H., Hamada, H., Noguchi, S., Kamada, Y. and Muramatsu, T. (1985) *Dev. Biol.*, **110**, 284–96

7 Interaction between cell surface and nucleus

As has been described, many cell surface molecules are involved in the regulation of cell differentiation. Besides cell surface molecules, those in the cell nucleus, especially DNA binding proteins, have been shown to play decisive roles in the regulation of differentiation and development. In this chapter mention will be made briefly of surface–nucleus interaction and of some DNA binding proteins in order to provide a more general view of the mechanism of differentiation.

7.1 DETERMINATION OF THE DORSAL–VENTRAL AXIS

The dorsal–ventral axis and anterior–posterior axis are determined during the early stages of embryogensis; these two axes provide information to a cell at a given position about the direction of its differentiation. The mechanism of axis determination is best demonstrated in *Drosophila* [1].

The dorsal–ventral axis is determined and is established as a result of complex molecular interactions just around the time the syncytial blastoderm is formed. Several maternal-effect genes are primarily involved in the determination process [2, 3]. Females carrying the null mutation of the genes produce eggs which develop abnormally. In most cases, the mutation is recessive and in the homozygous mutant, the ventral part shows dorsal characteristics (Figure 7.1). Some of the *Toll* mutants are dominant ones, and produce embryos with ventralized dorsal portions. This phenomenon can be interpreted as follows: The dorsal phenotype is the prototype on which the ventralizing factor acts to make ventral structures. The activity of the ventralizing factor probably makes a gradient along the dorsal–ventral axis and should be high in the ventral part. *Toll* mutation is considered to increase the ventralizing activity to abnormal levels.

The maternal genes appear to affect the embryonic phenotype through mRNA and proteins stored in the egg. The lethal phenotype of the homozygous embryos can be rescued by injection of cytoplasm or poly(A)$^+$ RNA isolated from wild-type embryos. By examining the localization of the rescuing activity and regional specificity of the injection sites, the maternal genes can be classified into 3 categories (Table 7.1). Anderson, Nüsslein-Volhard and co-workers [3] have proposed that several maternal genes

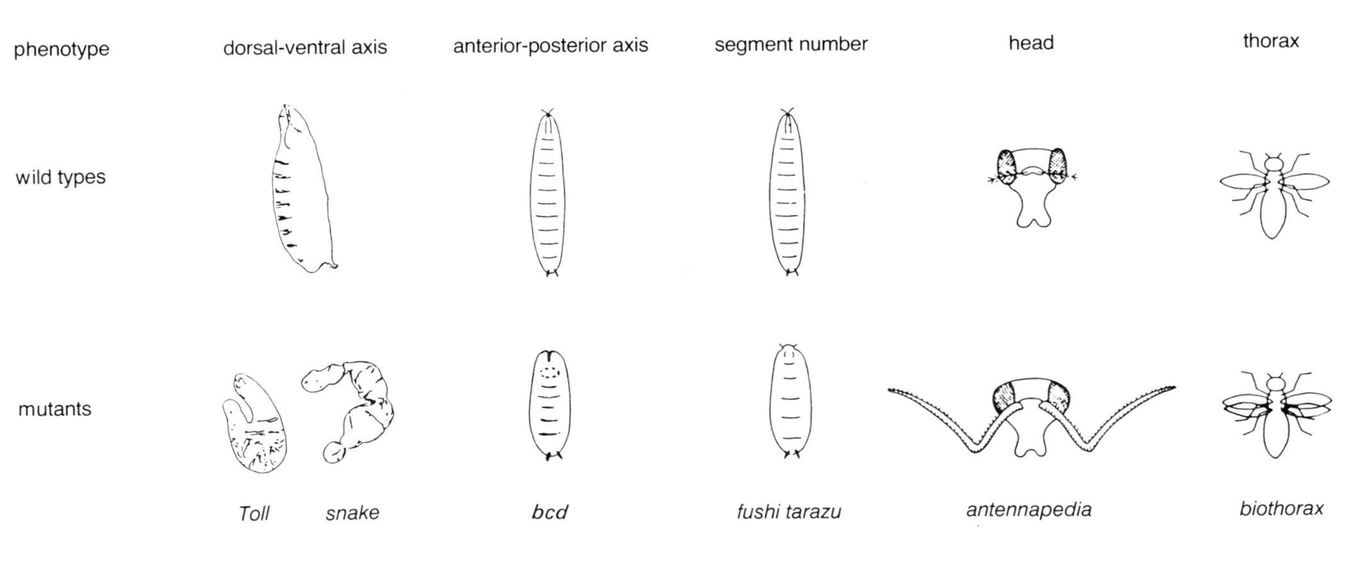

Figure 7.1 Phenotypes of *Drosophila* mutants causing abnormalities in embryogenesis.

Table 7.1 Classification of maternal genes involved in dorsal–ventral axis determination by rescue experiments

Mutants	Source of RNA or cytoplasm to be taken from wild-type embryos	Site of injection into mutant embryos
Snake	RNA present in any part of embryos was effective	Injection to any part of the mutant embryo rescused it
Toll	Cytoplasm from any part of the embryo was effective	Ventral structure is formed at the site of injection
dorsal	As the embryogenesis proceeded, the rescuing activity became to be more enriched in cytoplasm from the ventral region	Injection to ventral part of the embryo rescued it

involved in the determination of the dorsal–ventral axis operate through a cascade mechanism: *snake*, *pipe*, *nudel*, *easter* and *gastrulation defective* convert the *Toll* precursor to the *Toll* active form (Figure 7.2); the *Toll* active form then interacts with *dorsal* gene product to produce or activate the ventralizing factor. One of the key steps which generates dorsal–ventral polarity appears to be the activation stage of *Toll* gene product.

So far three genes, *snake*, *dorsal* and *Toll* have been cloned and their sequences have been established. *Snake* gene encodes a polypeptide of 430 amino acids. The C-terminal end, consisting of 246 amino acid residues, has homology (~ 30%) with serine proteases such as trypsin, chymotrypsin and thrombin [4]. The consensus sequences of the serine protease family, which form the active sites, are conserved in the *snake* gene product. Therefore the *snake* gene product appears to act as a protease. Thrombin and other molecules of the blood clotting system belong to the serine protease family. *Snake* gene product may be involved in a cascade of proteases, which finally yields an active molecule, just as in the case of the clotting system. In the N-terminal side, the predicted *snake* gene product has EF hand, which is the

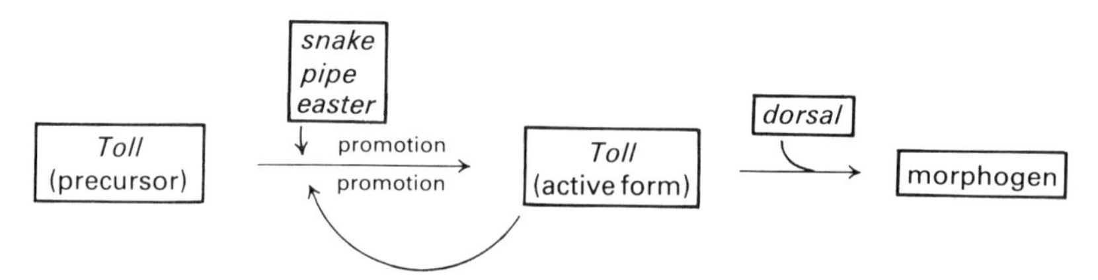

Figure 7.2 Postulated interaction between products of genes involved in the formation of the dorsal–ventral axis.

consensus sequence of calcium binding sites found in many proteins, suggesting that *snake* protein is regulated by calcium.

On the other hand, the *Toll* gene product appears to be a transmembrane polypeptide of 125 K [5]. At least 15 repeats of a 24 amino acid, leucine-rich, sequence have been detected in the extracellular portion of the *Toll* polypeptide (Figure 7.3). A leucine-rich sequence similar to that of *Toll* is present in chaoptin, a photoreceptor cell-specific membrane protein of molecular weight 127 K (as a polypeptide). Forty-one repeats of the leucine-rich sequence are present in chaoptin [6]. Mutation in the chaoptin sequence causes microvillar disorganization in developing rhabdomeres and disruption of the closely apposed membranes of adjacent cells. Chaoptin is thought to be involved in the adhesion between closely apposed membranes; the *Toll* gene product may also function in this way.

Dorsal gene codes a protein of molecular weight 75 K. The N-terminal half of the dorsal protein has about 50% homology with *rel* oncogene, (*v-rel* and its protooncogene, *c-rel*) [7]. *V-rel* is the viral oncogene of the reticuloendotheliosis virus strain T, and is highly oncogenic in avian lymphoid, spleen and bone marrow cells. Although *dorsal* mRNA is uniformly distributed in the embryo, *dorsal* protein is localized in the nuclei

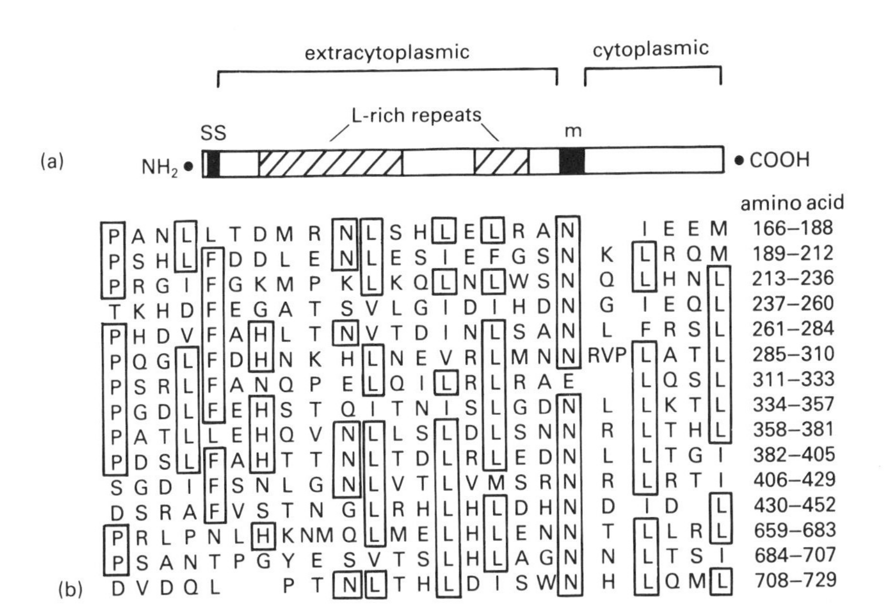

Figure 7.3 Structural organization of the predicted *Toll* protein (a) and the leucine-rich repeat (b). ■ hydrophobic region (ss: signal sequences; m: transmembrane segment). ▨ leucine-rich repeat. (Cited from [5].)

of the ventral region [8]. The level of *dorsal* protein is probably regulated at translation or by post-translational modification.

Toll polypeptide modified by a protease cascade involving *snake*, is the probable biochemical manifestation of active *Toll*. The decisive role of activated *Toll* in the generation of the axis and the fact that the *Toll* gene specifies a transmembrane protein suggest the importance of cell surface recognition in the generation of the dorsal–ventral axis. The cell-surface information acquired by the activated *Toll* is transmitted to the cell nucleus by activation or translocation of *dorsal* protein.

After the dorsal–ventral axis has been determined, this information seems to lead to the activation of a group of zygotic genes, which can be defined by respective mutants. Mutation in *Hin-d* cause defects in dorsal derivatives, and the embryo is ventralized. As has been described, the *Hin-d* gene encodes a factor similar to TGF-β. *Twist* and *snail* genes are required for mesoderm formation. *Single-minded* mutants have defects in the ventral ectoderm. *Snail* gene product is a DNA binding protein with a zinc finger [9], and *twist* also specifies a nuclear protein [10]. The product of *single-minded* (*sim*) gene [11] is similar to that of the *per* locus, which controls the periodicity of biological rhythms. Although the *per* gene product resembles proteoglycans [12], the *sim* gene product appears to be present in the nucleus [11]. Thus the realization of the dorsal–ventral axis by zygotic genes requires both cell-surface interactions, as in the case of *Hin-d*, and cascades of nuclear protein interactions, as in other cases.

7.2 DETERMINATION OF THE ANTERIOR–POSTERIOR AXIS AND PATTERN FORMATION

Molecular interactions yielding the head to tail (anterior–posterior) axis and pattern formation along the axis are being clarified rapidly [1]. The anterior–posterior axis has already been determined at the syncitial blastoderm stage. Three kinds of maternal genes are involved in the process; among them *bcd* is thought to be the major controlling gene. Mutant embryos devoid of *bcd* function lack the head and the thorax (Figure 7.1). When the cytoplasms of wild type embryos are transplanted into the *bcd* mutant embryo, a head structure is formed at the site of the injection. *bcd* RNA is localized only at the anterior pole of the oocyte, since it is synthesized by nurse cells and introduced into oocytes at the pole. *bcd* protein forms a gradient in the cleavage stage embryo with a maximum at the anterior tip, reaching background levels in the posterior third [13]. *bcd* gene has been cloned and the product shown to be a DNA binding protein – since *bcd* gene has a homeobox, which is found in several morphogenic genes and specifies a DNA binding region.

After the anterior–posterior axis has been established segmentation genes divide the embryo into segments. Three classes of segmentation genes are present, and all are zygotic genes [1]. Gap genes, such as *Krüppel* divide the embryo into large compartments. In the *Krüppel* mutant, the embryo lacks the thorax and the anterior portion of the abdomen. *Krüppel* gene product is a nuclear protein – it has a zinc finger, which is a motif of a class of DNA binding protein [14]. Pair rule genes are defined by mutants whose embryos have half the number of segments that wild-type embryos have (Figure 7.1): *fushi tarazu* [15], *evenskipped* [16] and *paired* [17] are examples of pair rule genes. They also specify nuclear proteins, since they have a homeobox. Segment polarity genes are defined by mutants whose embryos lack posterior structures in every segment; the posterior structures are replaced by the anterior structures duplicated in a mirror image pattern. *wingless* and *engrailed* are typical examples of these genes. *engrailed* [18] also has a homeobox, while *wingless* specifies a growth factor-like molecule as mentioned before.

The expression of gap genes is controlled by genes determining the anterior–posterior axis. On the other hand, gap genes control the expression of pair rule genes, which in turn appear to regulate segmentation polarity genes.

Each segment established then comes under the effects of homeotic genes which lead to the development of segment-specific features [19]. Abberent morphology occasionally results from the mutation of homeotic genes. In an *antennapedia* mutant, the antenna in the head portion are converted to legs (Figure 7.1). In a *bithorax* mutant, a thorax segment is altered so that the fly has four wings (Figure 7.1).

When the DNA sequence of *antennapedia* was compared with that of *bithorax*, portions of the genes were found to show intense homology [20, 21]. The DNA sequence was called a homeobox; the finding of a homeobox was a landmark in the molecular biology of development. A homeobox specifies a peptide domain, a homeodomain, which is rich in basic amino acids and has DNA binding capability. As mentioned above homeoboxes have been found in several genes involved in the morphogenesis of *Drosophila*. Furthermore, homeobox-containing genes have been detected in many organisms including humans. Developmentally controlled expression of homeobox containing genes in vertebrates implies that they also play important roles in vertebrate embryogenesis [22, 23, 24].

The expression of homeotic genes are controlled by polarity and segmentation genes [1]. Thus a series of molecular interactions, notably those of DNA binding proteins, regulates pattern formation along the anterior–posterior axis. Although molecular interactions within the nucleus are of primary importance in defining pattern formation, the realization of the pattern needs regulated expression of cell surface molecules. Therefore it is not unlikely that products of homeotic genes bind to and regulate the genes specifying certain cell surface molecules such as cell adhesion molecules. Indeed, a *trans*-acting factor required for B cell-specific expression of immunoglobulin genes has been shown to have a homeodomain [25].

7.3 RETINOIC ACID AS A MORPHOGEN

DNA binding protein is also important in the case of the determination of the polarity of digit formation. In limb morphogenesis, proximo–distal specification is controlled by a layer of mesoderm which is called the polarizing zone and is located near the tip of the limb bud. When the polarizing zone is grafted to another limb bud in the opposite position, double fingers with mirror image orientation are produced. The active principle in the polarizing zone has been identified as retinoic acid [26, 27]. Retinoic acid administered to a specific position of the limb bud exerts the effect identical to the grafted polarizing zone. The concentration of retinoic acid is indeed higher in the polarizing zone, and is sufficient to cause the effect. Within cells, retinoic acid is bound to receptor proteins homologous to steroid hormone receptors and the thyroid hormone receptor [28, 29]. The retinoic acid–receptor complex is expected turn on or turn off specific genes by binding to their promotor region. What genes are turned on by the receptor binding? In EC cells treated with retinoic acid, *int-2* (a member of the FGF family) is expressed in the earlier stages of differentiation [30]; concomitantly another member of FGF family, *hst* ceases its expression [31]. Thus, growth factors and possibly their receptors may be amongst the important molecules whose expression is regulated by retinoic acid. A morphogenic substance of cellular slime mold, which is required for the formation of the prestalk/prespore structure is also a non-polar material [32] and may work in a way similar to retinoic acid.

7.4 A MORPHOGENIC SUBSTANCE FOUND IN HYDRA

A morphogenic substance found in hydra is not a non-polar material, but a polar material of low molecular weight [33]. When a hydra is cut into pieces, the head (hypostoma) is regenerated from the upper end of the piece, and the basal disc is regenerated from the lower end. Grafting experiments have lead to the conclusion that head formation is governed by the gradient distribution of two substances, a head activator and a head inhibitor, along the body column of a hydra. The head inhibitor is a soluble molecule produced by the head. The intercellular gradient formed by a water-soluble, low molecular weight substance is expected to require coupling of the cytoplasm by the gap junction. Indeed, the epithelial cells of hydra are extensively linked by gap junctions. Antibodies against gap junction proteins inhibit slightly, but significantly, the action of the head inhibitor [34]. Gap junctions are known to close and open according to external signals. Therefore, modulation of gap junction opening and closure could be a means of effecting cell differentiation through cell surface interactions. The mechanism of action of the head inhibitor is not known, although the possibility is not excluded that the molecule, even though it is hydrophilic, is complexed with a receptor and regulates gene expression in a way similar to retinoic acid.

7.5 COMMENTS

One important question is how cytoplasmic signals, which are received and transferred into the cell by cell surface receptors, are transferred to the nucleus and regulate gene expression. This question is vital to understanding the regulation of various cellular activities including differentiation, but the answer is only just beginning to be understood. Recent evidence indicates that the phosphorylation of nuclear proteins is important [35]. Either proteinkinases themselves or their regulators may enter into the nucleus to change the state of protein phosphorylation. The presence of a DNA binding motif in proteinkinase C is intriguing in this respect. Glycosylation of transcriptional factors might also contribute to the control of gene expression [36]. Most probably several devices are present to control gene expression by cytoplasmic signals.

As we have seen, both cell surface molecules and nuclear molecules are of critical importance in regulating differentiation. Furthermore, both routes for the flow of information, namely from cell surface to nucleus and from nucleus to cell surface, are important in regulating and realizing differentiation. Stimulation of *c-fos* [37] and *c-jun* [38] expression by growth factors, the possible regulation of *dorsal* activity by activated *Toll*, and of homeotic genes by *wingless* gene product are suitable examples of surface–nuclear interaction where detailed analysis is expected to yield exciting findings.

REFERENCES

1. Ingham, P.W. (1988) *Nature*, **335**, 25–34
2. Anderson, K.V., Jürgens, G. and Nüsslein-Volhard, C. (1985) *Cell*, **42**, 779–89
3. Anderson, K.V., Bokla, L. and Nüsslein-Volhard, C. (1985) *Cell*, **42**, 791–8
4. DeLotto, R. and Spierer, P. (1986) *Nature*, **323**, 688–92
5. Hashimoto, C., Hudson, K.L. and Anderson, K.V. (1988) *Cell*, **52**, 269–79
6. Reinke, R., Krantz, D.E., Yen, D. and Zipursky, S.L. (1988) *Cell*, **52**, 291–301
7. Steward, R. (1987) *Science*, **238**, 692–4
8. Steward, R., Zusman, S.B., Huang, L.H. and Schedl, P. (1988) *Cell*, **55**, 487–95

9. Boulay, J.L., Dennefeld, C. and Alberga, A. (1987) *Nature*, **330**, 395–8
10. Thisse, B., Sstoetzel, C., Gotostiza-thisse, C. and Perrin-Schmit, F. (1988) *EMBO J.*, **7**, 2175–84
11. Crews, S.T., Thomas, J.B. and Goodman, C.S. (1988) *Cell*, **52**, 143–51
12. Jackson, F.R., Bargiello, T.A., Yun, S.H. and Young, M.W. (1986) *Nature*, **320**, 185–8
13. Driever, W. and Nüsslein-Volhard, C. (1988) *Cell*, **54**, 83–93
14. Rosenberg, U.B., Schröder, C., Preiss, A., Kienlin, A., Côté, S. *et al.* (1986) *Nature*, **319**, 336–9
15. Kuroiwa, A., Haften, E. and Gehring, W.J. (1984) *Cell*, **37**, 825–31
16. Frasch, M., Hoey, T., Rushlow, C., Doyle, H. and Levine, M. (1987) *EMBO J.*, **6**, 749–59
17. Frigerio, G., Burri, M., Bopp, D., Baumgartner, S. and Noll, M. (1986) *Cell*, **47**, 735–46
18. Fjose, A., McGinnis, W.J. and Gehring, W.J. (1985) *Nature*, **313**, 284–9
19. Lewis, E.B. (1978) *Nature*, **276**, 565–70
20. McGinnis, W., Garber, R.L., Wirz, J., Kuroiwa, A. and Gehring, W.J. (1984) *Cell*, **37**, 403–8
21. Scott, M.P. and Weiner, A.J. (1984) *Proc. Natl. Acad. Sci.*, **81**, 4115–9
22. Colberg-Poley, A.M., Voss, S.D., Chowdhury, K., Stewart, C.L., Wagner, E.F. and Gruss, P. (1985) *Cell*, **43**, 39–45
23. Hauser, C.A., Joyner, A.L., Klein, R.D., Learned, T.K., Martin, G.R. and Tijan, R. (1985) *Cell*, **43**, 19–28
24. Awgulewitsch, A., Utset, M.F., Hart, C.P., McGinnis, W. and Ruddle, F.H. (1986) *Nature*, **320**, 328–35
25. Ko, H., Fast, P., McBride, W. and Staudt, L.M. (1988) *Cell*, **55**, 135–44
26. Tickle, C., Alberts, B.M., Lee, J. and Wolpert, L. (1982) *Nature*, **296**, 564–5
27. Thaller, C. and Eichele, G. (1987) *Nature*, **327**, 625–8
28. Petkovich, M., Brand, N.J., Krust, A. and Chambon, P. (1987) *Nature*, **330**, 444–50
29. Giguere, V., Ong, E.S., Segui, P. and Evans, R.M. (1987) *Nature*, **330**, 624–9
30. Jakobovits, A., Shackleford, G.M., Varmus, H.E. and Martin, G.R. (1986) *Proc. Natl. Acad. Sci.*, **83**, 7806–10
31. Yoshida, T., Muramatsu, H., Muramatsu, T., Sakamoto, H., Katoh, O., Sugimura, T. and Terada, M. (1988) *Biochem. Biophys. Res. Commun.*, **157**, 618–25
32. Morris, H.R., Taylor, G.W., Masento, M.S., Jermyn, K.A. and Kay, R.R. (1987) *Nature*, **328**, 811–4
33. Berking, S. (1979) *Wilhelm Roux's Arch. Dev. Biol.*, **186**, 189–210
34. Fraser, S.E., Green, C.R., Bode, H.R. and Gilula, N.B. (1987) *Science*, **237**, 49–55
35. Yamamoto, K.K., Gonzalez, G.A., Biggs III, W.H. and Montminy, M.R. (1988) *Nature*, **334**, 494–8
36. Jackson, S.P. and Tjian, R. (1988) *Cell*, **55**, 125–33
37. Lau, L.F. and Nathans, D. (1987) *Proc. Natl. Acad. Sci.*, **84**, 1182–6
38. Quantin, B. and Breathnach, R. (1988) *Nature*, **334**, 538–9

Index